Rendered Obsolete

Rendered Obsolete

Energy Culture and the Afterlife of US Whaling

JAMIE L. JONES

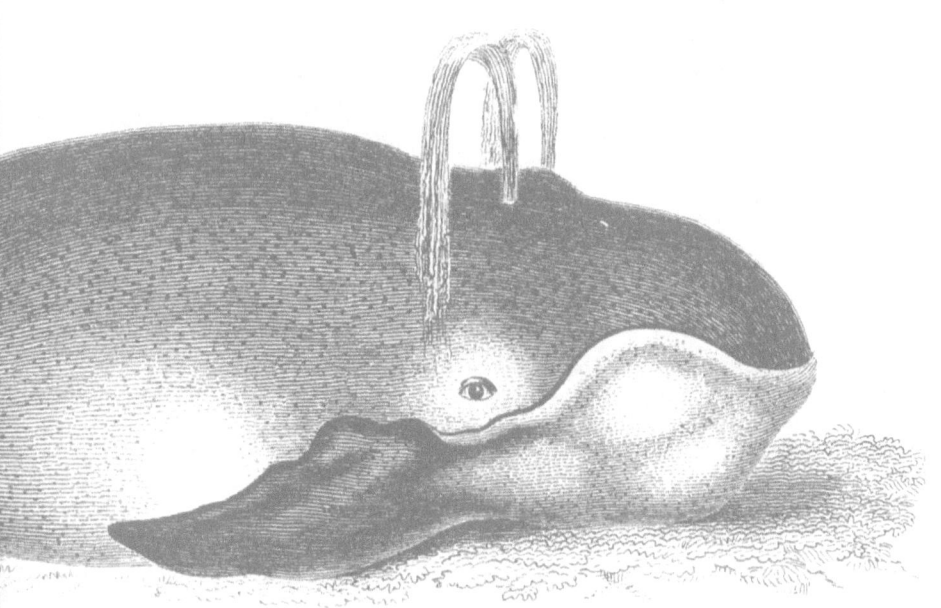

THE UNIVERSITY OF NORTH CAROLINA PRESS

Chapel Hill

This book was published with the assistance of the Wells Fargo Fund for Excellence of the University of North Carolina Press.

© 2023 Jamie L. Jones
All rights reserved
Set in Miller, Antique No 6, and Sentinel
by codeMantra
Manufactured in the United States of America

Library of Congress Cataloging-in-Publication Data
Names: Jones, Jamie L., author.
Title: Rendered obsolete : energy culture and the afterlife of US whaling / Jamie L. Jones.
Description: Chapel Hill : University of North Carolina Press, [2023] | Includes bibliographical references and index.
Identifiers: LCCN 2023010440 | ISBN 9781469674810 (cloth ; alk. paper) | ISBN 9781469674827 (paperback) | ISBN 9781469674834 (ebook)
Subjects: LCSH: Melville, Herman, 1819–1891. Moby Dick. | Whaling—Social aspects—United States—History—19th century. | Whaling—Social aspects—United States—History—20th century. | Whaling—United States—In popular culture. | Whale oil—United States—History. | Petroleum industry and trade—Social aspects—United States. | Petroleum industry and trade—Environmental aspects—United States. | Power resources—Social aspects—United States. | Power resources—Environmental aspects—United States. | White nationalism—United States—History.
Classification: LCC SH383.2 .J664 2023 | DDC 338.3/720973—dc23/eng/20230512
LC record available at https://lccn.loc.gov/2023010440

for Eric

and my parents, Connell and Karen Jones,

with love

CONTENTS

List of Illustrations ix

Acknowledgments xi

INTRODUCTION. UNDERGROUND WHALES
An Energy Archaeology 1

Part 1. LOOMINGS

1. BUILT-IN OBSOLESCENCE
Energy and Limits to Growth in the Whaling World of *Moby-Dick* 29

Part 2. WHALING ENTERTAINMENT

2. THE INVENTION OF QUAINTNESS
Nantucket Tourism and the Logics of Energy and Exhaustion 61

3. PIONEER INLAND WHALING
A Whale on a Train, a Ship Called *Progress*, and the Transformation of Whaling Culture in the Inland United States 93

Part 3. WHALING NOSTALGIA

4. EXTINCTION BURST
White Supremacy and Yankee Whaling Heritage at the End of the Industry 117

5. NOSTALGIA FOR THE WOODEN WORLD
Energy, the Melville Revival, and Rockwell Kent's *Moby-Dick* 150

EPILOGUE. THE BONE IN OUR TEETH 183

Notes 195

Bibliography 219

Index 237

ILLUSTRATIONS

Cartoon of "Grand Ball Given by the Whales in Honor of the Discovery of the Oil Wells in Pennsylvania," *Vanity Fair*, April 20, 1861   3

Photograph of steamboat passengers encountering obsolescing whaling ships in late nineteenth-century New Bedford, Massachusetts   72

Portrait of a quaint old ship's captain in Nantucket by Henry S. Wyer in *Nantucket Characters*   80

Promotional material advertising the exhibition of a whale's corpse in the 1880s   107

Photograph of the *Progress*, a whaling ship exhibited at the Columbian Exposition in Chicago   110

Photograph of the whaleman statue in New Bedford, Massachusetts   125

The Sea Chest, a painting by Clifford Ashley   138

A scrimshawed whale's tooth carved by Raymond McKee, star of the 1922 silent film *Down to the Sea in Ships*   140

Illustration of Captain Ahab by Rockwell Kent in the 1930 Lakeside Press edition of *Moby-Dick*   160

Colophon for Viking Press by Rockwell Kent   162

Fire Wood, an illustration Rockwell Kent made for his memoir *Wilderness: A Journal of Quiet Adventure in Alaska*   166

Rockwell Kent, a parodic portrait of the artist by Miguel Covarrubias   172

Toilers of the Sea, a painting by Rockwell Kent   176

Night Watch, an engraving made by Rockwell Kent to advertise motor yachts produced by the American Car and Foundry Company   177

Hail and Farewell, an engraving made by Rockwell Kent to advertise motor yachts produced by the American Car and Foundry Company   178

ACKNOWLEDGMENTS

This book is an expression of exultation, as Emily Dickinson defined it: "the going of an inland soul to sea." The crises of fossil modernity that this book describes are grave, but the relationships that sustained my research have given me great joy and profound gratitude.

I owe so much to the museums that steward, share, care for, and critique the history of the sea: Mystic Seaport Museum, the New Bedford Whaling Museum, and the Nantucket Historical Association. I owe countless of the insights in this book to these museums' curators, archivists, staff, and demonstration squad, past and present. Over the years, the curators at these museums have also invited me to make meaning with them as a public speaker, consultant, and consulting curator. I have been humbled by their trust, and inspired by the solemn task of stewarding history. Mystic Seaport Museum in Mystic, Connecticut, has been home (sometimes literally) for years. For their astonishing expertise in all matters maritime and support over the years, I owe special gratitude to Nicholas Bell, Mary K. Bercaw Edwards, Fred Calabretta, Craig Edwards, Elysa Engelman, Susan Funk, Katharine Mead, Paul O'Pecko, Maribeth Quinlan, Krystal Rose, Jonathan Shay, Quentin Snediker, and Stephen White. I am profoundly grateful, too, for Mystic Seaport for restoring and sailing the *Charles W. Morgan*, the last extant commercial wooden whaling vessel in the world, and for naming me a Voyager in the 38th Voyage of the *Charles W. Morgan*. Mystic Seaport Museum was my home in 2012 when I attended the Munson Institute, a NEH Summer Faculty Institute. I am grateful to Directors Glenn Gordinier and Eric Roorda; visiting faculty, including Jeffrey Bolster, James Carlton, Lisa Norling, Marcus Rediker, and Helen Rozwadowski; and my fellow classmates.

I have come to feel at home, too, at the New Bedford Whaling Museum, where I gave my first public talk on whaling history as a graduate student and where I have returned over the years to research, speak, and luxuriate in the curators' unmatched expertise and the museum's amazing collections. I owe special thanks to the past and present curators, staff, and scholars I have met there over the years, including Akeia Benard, Christina Connett Brophy, Michael Dyer, Stuart Frank, Judith Lund, Mary Malloy, and Mark Procknik.

As Herman Melville wrote, Nantucket is the "great original" in the business of US whaling, and the research I undertook at the Nantucket Historical Association and Whaling Museum have been indispensable to this project. Dan Elias, Mary Emery Lacoursiere, and James Russell have supported this work and invited me to the island. My presentation at the symposium honoring Melville's two-hundredth birthday alongside Mary K. Bercaw Edwards and Nathaniel Philbrick was a moment of special significance in my career. I owe particular thanks to the expert advice and indispensable support of Michael Harrison, Ralph Henke, and Amelia Holmes, who helped me navigate the archives of the NHA and shape my research questions.

I am also grateful to the brilliant folks at the Chipstone Foundation: Jonathan Prown, Sarah Carter, and Natalie Wright. They have welcomed me into the vital museum and art cultures they are creating on the third coast and around the world, and they always give my imagination full rein. Sarah Carter's name belongs in half a dozen places in this list; she is a visionary scholar, curator, and community builder, and I am so proud that she is my friend.

A number of other institutions have supported this research through fellowships and research opportunities. A predoctoral fellowship at the Smithsonian American Art Museum provided me with time, space, resources, and community to develop the art historical and visual culture dimensions of my work. I am forever grateful to SAAM curators William Truettner and Eleanor Jones Harvey, to fellowship coordinator Amelia Goerlitz, and to my fellow fellows: Jobyl Boone, Sarah Carter, Ellery Foutch, Holly Goldstein, Valerie Hellstein, Annemarie Voss Johnson, Jason LaFountain, Alex Mann, Jonathan Waltz, Melissa Warak, and Mary Beth Zundo. I am also grateful for the archivists at the Archives of American Art, where I researched during my fellowship. I am grateful to the John Carter Brown Library for the gift of the Marie L. and William R. Hartland Fellowship and time to research in their magnificent collections. The Charles Warren Center in American History provided time, space, and funding through a predoctoral fellowship, and I am grateful to the Whiting Dissertation Completion Fellowship for support at a crucial moment in graduate school.

I have always drawn energy (the best, most sustainable kind) from conversation with distant colleagues at conferences and invited talks. I am particularly grateful for the opportunity to have worked out methodological issues in the environmental humanities, environmental media, and material culture in panels at C19, MLA, ASA, and MMLA (The Civil War Caucus) together with John Levi Barnard, Colleen Boggs, Alenda Chang,

Stephanie Foote, Christina Gerhardt, Teresa Goddu, Jennifer James, Kyla Schuller, Ana Schwartz, and many other fellow panelists. Dana Luciano, in addition to writing some of the field's most vital provocations, has done the work of convening many of these conversations, and I am profoundly grateful that she has invited me into them. I cannot offer enough thanks, too, to the always marvelous Hester Blum without whose research in oceanic studies or vital mentorship this book and many others would not exist.

I have found great community in the energy humanities, thanks to vibrant seminars and conference panels at ACLA and SLSA, and to conveners and fellow panelists, including Stacey Balkan, Siobhan Carol, Olivia Chen, Jeff Diamanti, Corbin Hiday, Devin Griffiths, Joya John, Graeme MacDonald, Monica Mohseni, Michael Rubenstein, Mark Simpson, Michael Tondre, and Jennifer Wenzel. I owe special thanks to Jeffrey Insko for being a constant interlocutor, mentor, and friend in energy humanities and Melville studies.

A number of institutions and collectives have invited me to present this work-in-progress over the years, and I am grateful to them for the opportunity and vital conversation. I owe special thanks to my colleagues in Lisbon, including Margarida Vale de Gato, Cecilia Martins, and Edgardo Medeiros da Silva, for hosting me at the *Over Seas* conference in 2019. And I am grateful, too, to John Palmesino and Ann-Sofi Rönnskog of Territorial Agency for inviting me to the Chicago Architecture Biennial to think together with them about the Museum of Oil. A presentation to the Kaplan Environmental Humanities workshop at Northwestern University gave me new purchase on material in chapter 3, and I am particularly grateful to Lydia Barnett, Harris Feinsod, and Hi'ilei Hobart for their feedback. The Vcologies working group has been welcoming, and I am so glad to be in conversation with Elizabeth Miller and others.

I count many, many others in my far-flung intellectual communities, and wish to express profound thanks here to Jonas Akins, Stacy Alaimo, Paul Erickson, Rebecca Evans, Stephanie LeMenager, Jason Mancini, Sarah Mesle, Angela Miller, Alexander Nemerov, Jason Smith, Timothy Walker, Marina Wells, Jake Wien, and many others.

I have completed the work for *Rendered Obsolete* while studying and working at Harvard University, the University of Michigan, and the University of Illinois, and I owe much gratitude to mentors, friends, and colleagues in each of these places. At Harvard University, I learned how to make my way in academia and American studies with the help of fellow students who have become enduring friends and colleagues. The long list includes Christina Adkins, Sarah Carter, Mark Hanna, Jared Hickman,

Brian Hochman, Eve Mayer, and Brian McCammack. I am also profoundly grateful to many faculty members for their research, teaching, and models of collegiality. These faculty include Patricia Bellanca, Lawrence Buell, Joyce Chaplin, Nancy Cott, Jill Lepore, Louis Menand, Jennifer Roberts, Werner Sollors, and John Stauffer. I owe a particular and enduring debt of gratitude to Larry Buell for his work in the field of ecocriticism and for the example of his dedicated mentorship and inspired pedagogy. I hear his words in my head almost every day.

At the University of Michigan, my book project and I benefited from the support of colleagues and friends in the Department of English, the Sweetland Center for writing, and the Lloyd Hall Scholars Program. I am grateful to Scotti Parrish for indelible writing advice and mentorship, and to many colleagues for valuable feedback, support, and friendship. These include Gina Brandolino, Lily Cox-Richard, Clare Croft, Anne Gere, Anita Gonzalez, T Hetzel, Petra Kuppers, Karla Mallette, Laura Miles, Carol Tell Morse, Liliana Naydan, Michael Schoenfeldt, Nick Valvo, and Patricia Yaeger.

I could not have brought this book project to fruition in a more ideal community than that I have found at the University of Illinois at Urbana–Champaign. Many units on campus have supported my work and me: the Humanities Research Institute under the leadership of the awe-inspiring Antoinette Burton; the Center for Advanced Study under the leadership of Tamer Başar and Masumi Iriye and where I enjoyed the honor of a Faculty Fellowship; the Institute for Sustainability, Energy, and the Environment; and the Unit for Criticism and Interpretive Theory. I have found rich and rigorous intellectual exchange, as well as indispensable feedback on my writing, in the HRI Animal Turn Research Cluster that I organize with my collaborator and friend Jane Desmond; the IPRH-Mellon Environmental Humanities Working Group; the HRI Environmental Humanities Research Cluster; and the Americanist Workshop. I received incisive and caring feedback on draft chapters from many colleagues and friends in these communities, including Leah Aronowsky, John Levi Barnard, Clara Bosak-Schroeder, Chip Burkhardt, Antoinette Burton, Lucinda Cole, Jenny Davis, Jane Desmond, Virginia Dominguez, Carolyn Fornoff, Janice Harrington, Bob Morrissey, Ramón Soto-Crespo, Derrick Spires, Eleanora Stoppino, Rebecca Oh, and Pollyanna Rhee.

I do not know how to begin to thank my mentors in the English department: Stephanie Foote, Christopher Freeburg, Gordon Hutner, Robert Markley, Bruce Michelson, Justine Murison, Robert Dale Parker, Curtis Perry, Michael Rothberg, Derrick Spires, Renée Trilling, and Gillen D'Arcy Wood. They have advocated for me, read countless drafts of countless projects, given me opportunities, checked in on me, and provided encouragement. This book owes its existence to the support of many other colleagues and friends in English and throughout the university, as well: Robert Barrett, Dale Bauer, Jayne Burkhardt, Nancy Castro, Kate Clancy, Lucinda Cole, Eleanor Courtemanche, Tim Dean, Adam Doskey, Stephanie Foote, Samantha Frost, John Gallagher, Jessica Greenberg, Amy Hassinger, Elizabeth Hoiem, Irvin Hunt, Lilya Kaganovsky, Brett Kaplan, Susan Koshy, Melissa Littlefield, Trish Loughran, Vicki Mahaffey, Lori Newcomb, Tim Newcomb, Amy Powell, Catherine Prendergast, Allyson Purpura, Dana Rabin, Kristin Romberg, Emanuel Rota, Lindsay Rose Russell, Robert Rushing, Ted Underwood, Deke Weaver, and Terri Weissman. I am also grateful to the research support wizards at the Office of the Vice Chancellor for Research and Innovation, who have helped me pursue grant opportunities and learn how to explain my project to others (and myself). They include Craig Koslofsky, Cynthia Oliver, Gabriel Solis, Carol Symes, and the fabulous Maria Gillombardo.

I also owe profound thanks to my students: undergraduate and graduate at Harvard, Michigan, and Illinois. You teach me every single week how to make knowledge together in community. For working alongside me in the classroom and as interlocutors in literary studies and environmental humanities, I owe special thanks to Leah Becker, Yoonsuh Kim, Jessica Landau, Lilah Leopold, Miya Moriwaki, Daniel Myers, Alexis Schmidt, Dana Smith, Stephanie Svarz, Carl Thompson, and the brilliant students in English 547.

I have many people to thank for making this project into an actual book, first and foremost my brilliant editor, Lucas Church. Thank you for believing in this project, Lucas, and for shepherding it into existence. Two anonymous readers provided me with crucial feedback and suggestions for revision; and I wish to thank them profusely for their care and labor. I am grateful to the whole dazzling team at the University of North Carolina Press: Thomas Bedenbaugh, Alyssa Brown, Valerie Burton, Laura Jones Dooley, Andreina Fernandez, Elizabeth Orange, Rebecca Rivette, Lindsay Starr, and others. Thanks are also due to my wonderful indexer, Amron Lehte.

I am also grateful to the artists, artist's estates, and institutions that have helped with images and permissions. For permissions to reprint work by Rockwell Kent, I thank Ceil Esposito, Tonya Cribb, and Karen Blough; for permission to reprint the work of Miguel Covarrubias, I thank María Elena Rico Covarrubias; and for help acquiring images for this book, I am deeply grateful to Amelia Holmes at the Nantucket Historical Association; Paul O'Pecko, Maribeth Quinlan, and Claudia Triggs at Mystic Seaport

Museum; Mark Procknik at the New Bedford Whaling Museum; James Kohler at the Cleveland Art Museum; Keith Gervase at the Cleveland Art Museum; and Adam Doskey and Lynne Thomas at the Rare Book & Manuscript Library at UIUC.

Portions of this research have appeared before in the following publications: "Print Nostalgia: Skeuomorphism and Rockwell Kent's Woodblock Style." *American Art* 31, no. 3 (Fall 2017); and "Fish out of Water: The 'Prince of Whales' Sideshow and the Environmental Humanities," *Configurations* 25, no. 2 (Spring 2017). I thank the editors of these journals for their editorial insight.

For weathering pre-tenure life and work together and for saving my sanity at least once a week for the past several years, I need to thank three brilliant fellow scholars: Na'ama Shenhav, Heather Soyka, and Laura Voith, the group formerly known as my FSP small group and forever known as my comrades.

Many friends, near and far, supported this work over the years. I owe sincere thanks and love to Alison Anders, Rebecca Curtiss, Skip Curtiss, Sam Erman, Catharine Fairbairn, Karen Frazier, Charlie Frazier, Ellen Goodman, Jonathan Hafer, Avital Livny, Catherine Malmberg, John Meyers, Ben Miller, Mark Palmenter, Elisabeth Pollock, Greg Pollock, Rachel Pollock, Christian Ray, Doug Steen, and Jonathan Tomkin. It is one of the great joys of my life that so many of the people listed elsewhere in these acknowledgments are friends, too, and that so many friends listed here are also vital interlocutors. I am grateful for all of you.

My parents, Connell and Karen Jones, gave me everything. They believed fiercely in my education, and they fought for it and for me. They taught me to be curious and ambitious, and they took me to the ocean, giving me my first glimpse of exultation. I have the best sister in the world, Emily Jones, and I could not be luckier that she is my colleague and sister-friend, too. I thank her for bringing Daniel Le Ray into our lives, and Bennett and Fionn, too. I love you all very much.

I have the best luck in in-laws, and I could not be more grateful to Stephen and Nancy Calderwood for supporting my husband and me throughout our careers, and for Michael, Audrey, Sam, and Katie Calderwood for welcoming me into your family and lives.

My deepest thanks go to my husband and partner in everything, Eric Calderwood. You have believed in this work and supported me in ways I did not even know were possible. You talked me through all my happy discoveries and worried nights. You are my ideal reader and dream shipmate. Thank you for the adventures, the spacious days, and the great love.

Rendered Obsolete

INTRODUCTION

Underground Whales
An Energy Archaeology

The Pennsylvania oil fields were full of whales. Reporters at the site of the United States' first oil boom in the 1860s wrote that oil wells "spouted" like whales coming up for air, and oil shot through pipes from the well to the holding tank "with a sound like the 'blowing' of a whale."[1] According to a Kansas newspaper reporter in 1861, some oil field workers even imagined that underground oil reserves were fossilized whales: "Some of the Philosophers think this country was once an inland sea inhabited by the monsters of the deep, and that oil as found was the death bed of an antediluvian whale. Oil is imperishable, every vestige of the animal is gone but the grease."[2]

Perhaps oil field workers saw whales because so many of them had once worked in the whaling industry. And why wouldn't the whalemen go into petroleum? Oil was oil, and whalemen were oilmen. Before the development of those Pennsylvania oil fields, whaling had been the primary oil industry of the late eighteenth- and nineteenth-century United States; whale oil was used to light lamps and lubricate machinery.[3] Until the early 1860s, newspaper reports about the "oil industry" or the "price of oil" referred to whaling and whale oil. But as the new oil fields opened up, rock oil (petro + oleum) supplanted whale oil, and petroleum became the oil that mattered most. It is no surprise that the oilmen saw whales in Pennsylvania: such is the power of resource extraction to make one thing into another. The commercial whaling industry had made whales into oil, so it took only a short intellectual leap to believe that oil might be made of whales.

As the petroleum fields grew, oilmen from the whaling industry sought better opportunities in the new petroleum business. One observer, writing for the *New York Tribune*, observed the migration of workers from the whale fishery to the oil fields: "I find that New Bedford and Nantucket, heretofore oildom, has been unsuccessful for several years past, and is coming here, with its millions of money and its hordes of vessel officers, to harpoon the old mother of all whales (earth) and draw her blubber by the force of steam, which must eventually injure whaling oildom very much."[4] New England whalemen founded refineries, built tank infrastructure, and opened hotels for oil field laborers. In 1861, a whale oil refiner named Charles Ellis and H. H. Rogers, son of a Fairhaven, Massachusetts, whaleman, built a refinery near Oil City, Pennsylvania. Within the first year of the refinery's operation, "[Rogers's] cruise to Pennsylvania had netted him as much as half a dozen whales." Rogers later became a director at John D. Rockefeller's Standard Oil.[5] Back in New Bedford, Massachusetts, whaling magnate Abraham Howland converted a whale oil refinery over to kerosene and coal oil.[6] New Bedford whaling captain and shipowner John Arnold Macomber built and operated oil storage tanks in Pennsylvania.[7] And the Crape House hotel for workers in Oil City was founded by a New Bedford man "who, like others from the same quondam oily city, now follow oil wherever they can smell it."[8]

In 1861, *Vanity Fair* published a cartoon that dramatized this moment in the oil market from the point of view of the whales: "Grand Ball Given by the Whales in Honor of the Discovery of the Oil Wells in Pennsylvania." In the cartoon, sperm whales dressed in evening attire dance, sip cocktails, and pop bottles of champagne. They are attended by tiny frog-waiters in coattails in a ballroom strung with banners promoting Pennsylvania oil: "Oils Well That Ends Well," for example, and "We Wail No More for Our Blubber."

That cartoon of dancing whales may as well mark the invention of "energy" as a unified field of labor, production, and circulation based, not on the conditions of the resource's extraction, but on the consumption of its end product. Commercial whaling and commercial oil drilling demand wildly different forms of labor and infrastructure: murdering whales at close range and boiling their bodies in the middle of the ocean, in the case of the whaling industry, and drilling an oil well underground in the middle of the continent, in the case of petroleum. But the marketplace for lighting oil brought these disparate natural resources and extraction practices together within an idea called an oil market and a unified environmental imaginary that would be called energy.[9]

Whales and whaling were everywhere in accounts of the mid-nineteenth-century Pennsylvania oil boom, and they were symptoms of an emerging myth about energy: that energy was energy, wherever it came from. The idea of energy elided the difference between different fuel sources and gave rise to one of the enduring environmental myths of past few centuries: that energy sources are infinitely interchangeable because their productive power can be measured in standardized units such as calories, lumens, joules, or megawatts. The idea of energy erects a barrier between energy

The cartoon of dancing whales celebrating Pennsylvania oil shows how "energy" came to be understood in the mid-nineteenth century as a category independent of specific resources such as whale oil or petroleum. "Grand Ball Given by the Whales in Honor of the Discovery of the Oil Wells in Pennsylvania," cartoon, *Vanity Fair*, April 20, 1861. Courtesy of the New Bedford Whaling Museum.

production and consumption, and it collapses the difference between, say, whale oil and petroleum, wood and coal, fossil fuels and renewable energy sources. Even the tuxedoed whales of the *Vanity Fair* cartoon internalized the emerging logic of energy. They did not desperately advocate the end of whaling on behalf of their mortally endangered selves; instead, they saw themselves as interchangeable with petroleum and gave urbane champagne toasts to the smooth transition of energy from one source to another. In the story of this nineteenth-century transition from whale oil to rock oil, the concept of "energy" was born and, along with the concept, all of the opportunities and problems that characterize life with fossil fuels.

But the assumption that all energy sources are interchangeable is a myth, one of many myths about energy that this book seeks to name and unravel. In reality, different energy regimes make possible vastly different economies, technologies, infrastructures, sensory experiences, relationships, attachments, class and political structures, feelings, and social norms. It's not just that we use a fuel like petroleum; as literature and environmental studies scholar Stephanie LeMenager has suggested, we *live* oil. Through transportation, electricity, and materials such as plastic, fossil fuels are the medium through which all of modern US culture is produced.[10] Political structures develop in concert with different forms of energy, too: consider the geopolitics of energy extraction and the way people have to labor differently, and sustain different infrastructures, to produce coal, oil, gas, or solar power.[11] The differences in the way different energy regimes shape life and culture emerge at moments of what's called "transition," when one way of powering the world gives way to another.

Rendered Obsolete chronicles the massive energy transitions of the nineteenth century in the United States, when an energy regime predicated primarily on organic fuel sources such as whale oil, tallow, wood, and the labor of human and animal bodies began shifting to an energy regime predicated primarily on the extraction and consumption of fossil fuels. Energy is notoriously difficult to divide into neat periods. As critic Jennifer Wenzel has noted, "The oil era is also the coal era and, for millions around the globe, also the era of dung, wood, and charcoal."[12] In the nineteenth-century United States, the whale oil era is also the era of wind, coal, kerosene, phosphorus, beef tallow, wood, human and animal muscle, and so on.[13] It is necessary to be skeptical about such efforts of periodization and even of the concept of energy transition itself. Because energy resources are rarely, if ever, phased out completely, energy history might more productively be understood as a series of additions, rather than transitions.[14] But in the period that this book covers, from 1851 to 1930, the United States transitioned from mostly *organic* fuel sources to mostly *fossil* fuel sources, to amend a nomenclature productively developed by historian Christopher Jones.[15] Most histories of this broad nineteenth-century energy transition focus on the ascendent fossil fuels and their radical effects on US life and culture. But the story of new and old energy sources is more complicated than that.

This book follows one of these descending lines in the organic-to-fossil energy transition: the US commercial whaling industry. The main commodity of the industry was whale oil, which was used as a source of lighting and machine lubrication. As such, it became a key material component of urbanization and industrialization.[16] Whaling voyages also brought back other products for market: ambergris, used in perfumes and medications, and baleen, also known as whalebone, used to construct corsets and umbrellas. The ascendant petroleum industry delivered the US commercial whaling industry its first death blow, and the industry's end was hastened through the 1860s and 1870s by a series of calamities that destroyed crucial whaling infrastructure. The first catastrophe was the Civil War. Confederate raiders targeted and sank US whaling ships, interpreted rightly as engines of Union economic power.[17] The Union, too, destroyed whaling ships by requisitioning them for naval use.[18] The second catastrophe was climatic. In the early 1870s, even more whalers than usual were crushed by sea ice while wintering over during particularly cold, icy winters in the Pacific Arctic.[19] Through the end of the nineteenth century, some whalers shifted their base of operations to San Francisco and Hawai'i, and an even smaller and diminishing fleet operated out of New England ports. Commercial whaling in the United States was effectively dead by the 1920s.[20]

But the story of the US whaling industry is not one of simple disappearance. Between the years of the first petroleum boom and the departure of the last commercial whaling voyage in the 1920s, the whaling industry still dispatched ships, slaughtered whales, and delivered oil and baleen back to US ports, albeit less and less with every passing decade. The whaling industry remained a feature of popular culture in ways that did not abate as the industry itself declined; energy regimes have a long cultural afterlife.

Rendered Obsolete argues that whaling, during the industry's decline and obsolescence, profoundly shaped US culture, even as the United States was rapidly adopting fossil fuels and the technologies and forms of life they make possible.[21] Specifically, whaling culture mediated the organic-to-fossil fuel energy transition of the late nineteenth and early twentieth centuries. Whaling culture became a location from which to take a compass reading of energy and modernity itself. The whaling industry was the subject of a wide and multifarious body of cultural production: whalemen's autobiographies, novels, paintings, prints, and lurid newspaper accounts; traveling moving panorama performances and lyceum lectures; artifacts from the industry such as harpoons, wooden casks, boats, coiled ropes and wooden tubs, brutal lances and knives, and even entire ships that found homes in new museums; and the body parts of whales, too, from scrimshawed teeth and bones to entire whale corpses, all hauled up from the ocean and exhibited for landbound audiences sometimes thousands of miles from the sea. These texts, artifacts, and bodies told stories about the obsolescing whaling industry, and at the same time those stories made meaning about fossil-fueled modernity and the world to come. The language and imagery of whaling mediated the transition to fossil modernity, while at the same time the physical residue—the ships, tools, infrastructure—became the material for new experiences of tourism, museums, public space, books, and history in fossil modernity.

This book traces the persistence of the whaling industry in five sites of US culture: (1) literature, through Herman Melville's *Moby-Dick*, which provided, among other things, a description and critique of the emerging concept of energy; (2) tourism, especially in impoverished coastal towns whose industrial decline became "quaint"; (3) popular entertainment, and in particular traveling exhibitions that the fossil-fueled railroad made possible and that seemingly collapsed the time and distance between the coast and the enormous US inland; (4) public commemorations that indexed the racial anxieties that surrounded energy transition; and (5) literary history, in the Melville Revival and the early institutions of US literary studies that interpreted energy transition along with literature and culture.

This book does not—could not—chronicle the end of whaling in the world. Whaling has not ended. Through the end of the nineteenth century and into the mid-twentieth century—the period I am calling the US commercial industry's decline and death—global whaling actually accelerated as new technologies made it possible to slaughter whales on an even larger scale. As whale oil fell out of use for lighting purposes, it gained popularity as an animal fat in margarine, soap, and other products.[22] In 1946, the International Whaling Commission (IWC) seriously curtailed industrial whaling, and in 1982, it enacted a full moratorium on commercial whaling among its member nations. Even still, industrial whaling has not yet ended. A small fleet of commercial whaling ships from Norway, Iceland, and Japan operate in resistance to the IWC moratorium. And whaling for subsistence and cultural tradition is a vital part of life for Indigenous people whose whaling practices are not covered under the IWC moratorium.[23] Whaling continues as a vital form of subsistence and a sustainable environmental practice among Indigenous nations including the Iñupiat, Makah, Yupik, Chukchi, and Nuu-chah-nulth.[24] The practices of Indigenous whalers in the Arctic and industrial whalers elsewhere are not hermetically distinct from the practices of nineteenth-century US whaling as described in this book; it is important to recall that a huge number of Indigenous people from nations all over the world participated in the nineteenth-century US commercial whaling industry and that contact with US commercial whalers changed whaling technologies and practices among Indigenous nations.[25] Nevertheless, whaling in the United States considerably changed in the late nineteenth and early twentieth centuries. The US commercial fleet, which had been the most productive whaling fleet in the world in the nineteenth century, participated very little in commercial whaling after the nineteenth century. The US populace had other sources of oil for fuel and other sources of animal food in the ballooning livestock industry of the West. Unlike other whaling industries in other places and times, US commercial whaling was an oil industry and a proto-energy industry; that is what was said to have died in the mid-nineteenth century. Where did the whaling industry go when it died?

FOSSIL MODERNITY

At its most basic level this is a book about how whaling is implicated in the emergence of what I call fossil modernity. I use "fossil modernity" to denote a period of US and planetary history characterized by intensive fossil fuel consumption. The term also serves more expansively to describe the ways of work, leisure, life, culture, and identity associated with massive fossil fuel consumption. The growth and increased accessibility of transportation networks such as the railroad, steamship lines, and airlines are major features of fossil modernity, as is regular access to electricity and all of the mechanized and digital processes that electricity enables. In fossil modernity, fossil fuels become the pretext for ongoing settler colonialism in North America and US imperial incursions abroad. Fossil modernity encompasses new forms of cultural production and consumption: new media such as film, radio, television, computers, and digital media, and new ways of accessing and seeing old media such as books. Fossil modernity includes the feeling of driving a car with opened windows and of stepping into the air-conditioned climate of an airport's arrival lounge. Fossil modernity encompasses, too, the types of labor that fossil fuel extraction requires and the lives of laborers who work in coal mines, on oil rigs, on pipeline construction teams, and in refineries. Fossil modernity describes refinery towns with poisonous air, Appalachian towns whose mountaintops have been blown off, and Gulf Coast communities sinking, through subsidence and sea level rise, into the salty sea. The enormous acceleration of fossil fuel extraction and consumption in the late nineteenth and twentieth centuries is a fact, and it is a fact, too, that fossil fuels and the technologies they powered rewrote the US landscape at every scale, reshaped political and cultural forms, and generated new affects and attachments, many of which have been called modern. The concepts of modernity and the modern have long operated in colonial cultures as apologies for imperial domination and settler colonialism.[26] It is my hope that the term "fossil modernity" can help parochialize the term "modernity" by linking it to locally determined and historically contingent practices of fossil fuel extraction and consumption. While "fossil modernity" is in many ways an imperfect term, I use it here in an attempt to capture the material reality of increased fossil fuel consumption in the United States in the nineteenth and twentieth centuries and to name the ways of life fossil fuels sometimes, imperfectly and unevenly, make possible.

Fossil modernity is more of an idea than a period with clear beginnings and endings. Energy is heterochronous, resistant to periodization even as historians try to locate the beginnings and endings of the "ages" of oil, wood, coal, solar power. From the dancing whales of the *Vanity Fair* cartoon to twenty-first-century politicians imploring the world to embrace "renewables," we have always had a strong impulse to imagine history as a neat procession of energy sources: wood to coal to oil to nuclear energy to (our new hope) renewable energy. This book complicates that narrative by contributing to an account of what Wenzel has called the "untidiness and unevenness inherent to a history according to energy."[27] It is not just that different energy sources are in use at any given moment, frustrating a historian's desire to link eras with energy sources: even when energy sources fall out of use and into a state of real obsolescence, the forms of labor, technology, and culture they produced persist and continue to work. As this book's opening anecdotes demonstrate, the conceptual templates for oil extraction, refinement, and distribution developed in the whaling industry carried over into the Pennsylvania petroleum industry in the 1860s. And in the decades of the US commercial whale oil industry's long obsolescence through the late nineteenth and early twentieth centuries, the artifacts of the whaling industry, its infrastructures, and the stories told by its laborers continued to circulate and contribute meaning to the emerging idea of fossil modernity.

My term "fossil modernity" also evokes the persistence of the old in the new, the fossil in the modern. The case of the US whaling industry's long cultural afterlife shows how old energy and technological regimes, in addition to new fossil fueled ones, are integral parts of fossil modernity. This book offers a history of the idea of fossil modernity, a history of how it emerged, not as a fully formed new idea, but as a new configuration built precisely from those ideas that it called obsolete. As those early accounts of the Pennsylvania oil boom attest, whales and whaling culture permeated early petroleum culture. At the cusp of fossil modernity in the oil fields of Pennsylvania, it was impossible to understand petroleum without reference to the labor, language, and cultural logics of commercial whaling. Wooden whaling ships, harpoons, quaint New England port towns, whale oil lamps, salty old sailors: if these things sound old-fashioned or antimodern, it is because fossil modernity needed a foil. But the obsolescing remains of older energy regimes were not just a distant point of contrast, especially in the early days of the petroleum industry and before that industry had settled into its identity as "energy."

Obsolete energy regimes that work through new ones are an example of what critic Raymond Williams calls "residual" culture: "The residual, by definition, has been effectively formed in the past, but it is still active in the cultural process, not only and not at all as an element of the past, but as an effective element of the present."[28] In the period under discussion in this book, 1851–1930, whaling culture is a residual energy culture, a culture that remains present and operative even as fossil modernity sets in. Moreover, many of the texts, objects, and performances I chronicle here are what media scholar Charles Acland calls "residual media."[29] Residual media, as Acland argues, consist of "artifacts," "historical traces," or "things and sentiments ... that won't stay dead, lost, and buried."[30] Acland argues that residual media are "unambiguously a part of the everyday of modernity," much as I argue that old energy regimes are unambiguously a part of fossil modernity.[31] My terms "obsolete" and "fossil" closely follow Williams's and Acland's term "residual," but I use the terms "obsolete" and "obsolescing" in order to capture the colloquial usage of the term to describe technology culture, as well as the tension and irony of a supposedly dead industry that continues to work in the present. The whales and whalemen in the Pennsylvania oil fields of the 1860s were just the beginnings of the US whaling industry's work in fossil modernity. The decline of US whaling was not merely an effect of the massive fossil fuel consumption that kicked off in the mid-nineteenth century. The US whaling industry was one of the conditions through which fossil modernity emerged; whaling culture scaffolded fossil fuel culture. Fossil modernity is not a shiny alien spaceship that arrived from the future but a junkyard creature built from salvage.

The case of the whaling industry shows that conventional life-and-death metaphors do not adequately capture the complex and messy continuities between old and new industries, technologies, and marketplaces. *Rendered Obsolete* complicates the story of economic decline by tracking the emergence of new cultures at the sites of supposed decline. In so doing, this book revises the way we understand economic change. In 1942, political economist Joseph Schumpeter used the phrase "creative destruction" to describe the process by which capitalist economies create new markets in the act of destroying old ones.[32] "Creative destruction" applies to the mutually reinforcing destruction of the whale oil market and the creation of the petroleum market in the mid-nineteenth century, but that phrase only tells part of the story. "Creative destruction" evokes a tidy vanishing, but nothing ever vanishes. The story of the whaling industry in its decline and obsolescence shows how dead industries leave traces and continue to perform cultural work long after they have exhausted their original economic utility.

If the word "obsolete" is meant to emphasize persistence in the face of uselessness, "rendering" is doing similar work, evoking the multiple meanings of the word "render" explored by animal studies scholar Nicole Shukin.[33] "To render" means to break or boil down animal bodies, the process at the heart of the US whaling industry: whale bodies were rendered into oil. "Rendering" also means making a copy, an image, or a reproduction of something. The rendering of texts, images, and artifacts from and about the US whaling industry to a wide public was also one of the industry's central processes before and after its decline.[34] Both forms of rendering coincide in the whaling industry, making it an important site at which to reflect on both the material and the imaginative work of rendering during an energy industry's growth and well into its obsolescence. Shukin restores animality to the word "rendering," demonstrating the complicity of reproduction technologies with animal violence.[35] In evoking "rendering" in the title of this book, I am expanding the concept from the realm of animal life that Shukin describes to the extraction of such other resources as oil, coal, and minerals. I try to witness the violence of extraction, too, from nineteenth-century whaling to the climate-changed world of the twenty-first century. Whaling culture did not disappear in fossil modernity; whaling culture was rendered into fossil modernity.

IS WHALING ENERGY?

The persistence of whaling culture was not in contradiction with emergent fossil modernity; whaling culture actually supported that emergence by becoming obsolete. Stories about the US whaling industry during the years of its decline and obsolescence demonstrated the link between fuel sources and the infrastructures of travel, commerce, daily life, and even emotional experiences. Whaling and its artifacts became, through such cultural products as novels, exhibitions, artifacts, and films, metonyms for a pre-fossil-modern world. Precisely *because* whaling culture stood outside fossil modernity in the cultural imagination of energy, whaling culture helped explain how fossil fuels could and, eventually, did change everyday life in the fossil-modern United States.

But whaling has an uneasy history within the history of energy. Whaling was an oil industry, but it was not exactly an energy industry. "Energy," to be sure, is an anachronism for almost any fuel source produced in the nineteenth century. No one at the time called whaling an "energy industry." The nineteenth-century US whale oil industry operated on a different scale than petroleum eventually would; it was smaller and less politically powerful than oil, its applications were less extensive and varied, and its product, whale oil, was less ubiquitous in everyday life of the nineteenth century than petroleum would become in the twentieth and twenty-first centuries. In the mid-nineteenth century, the word "energy" did not even apply to coal, or to petroleum in the early days of the Pennsylvania oil boom, much less to whale oil, beef tallow, kerosene, or any of the other oils or illuminants in the marketplace.[36]

Sometime between the nineteenth and the twenty-first centuries, though, "energy" has come to mean "fuel," or that force which the consumption of fuel makes possible. Energy has become a *system* that, as critic Frederick Buell has argued, "came together through the dynamic and complex interaction of myriad factors: namely, energy sources, the technological arrangements necessary for their use, and the social arrangements that surround these material components."[37] It has also become a *sector* of the economy in the twenty-first century that encompasses coal, oil, and gas producers, the companies that build pipelines and dispatch trains and trucks, the power plants that consume energy to produce electricity, the gas stations that distribute refined oil to car-driving consumers, and the huge, networked assemblage of companies, infrastructures, services, and products that energy makes possible. Energy has also become a *field of scholarly research*, not only among engineers and geologists who support energy industries, but among historians, literary critics, and others working in the energy humanities, a subfield of the environmental humanities that explores the historical, social, and cultural dimensions of energy extraction, circulation, and consumption.[38]

"Energy," in this sense of the word, is a recent concept. Political scientist Cara New Daggett argues that "energy," as used in reference to fuels and fuel-infrastructure assemblages, emerged in the United States, Great Britain, and Europe in the late nineteenth century as a "ruling idea" that came to signify the physical force that does work, and the fuel sources that make it possible.[39] This sense of the term "energy" arose along with and *because of* the intensification and acceleration of fossil fuel consumption during the nineteenth century.[40] Before the term was linked to fuels in the mid- to late nineteenth century, "energy" was "a word predominantly for poets," a word to describe ineffable, ephemeral, or spiritual forces, a word that trod the line between metaphor and actuality.[41] The term got yoked to fuels and the concept of work through thermodynamic science, which itself arose, as Daggett argues, as a way to explain the workings of the steam engine.[42] Nineteenth-century thermodynamic science invested the word "energy" with a new conventional meaning—the ability to do work—which enabled the concept to become abstracted from working bodies and fuel sources into disembodied force. In fossil modernity, energy was increasingly supplied by fossil fuel sources that were more productive than human or animal laborers. The US whale oil industry does not figure into Daggett's account of the way energy became a ruling concept linked with work. "Energy," in this sense, is an anachronism when applied to whaling.

Other energy historians avoid the anachronism problem by deploying "energy" in its metabolic sense, not as a historically contingent concept linked with the acceleration of fossil fuels, but as an elemental force that derives ultimately from the sun. Energy, in the elemental sense, is a basic requirement for life in the form of calories, heat, and light.[43] Historian Bathsheba Demuth, for example, narrates the history of the Bering Strait from the standpoint of metabolic energy conversions, from solar radiation through photosynthesis in algae and plants and through animal tissue of ever-expanding scale up through fish, humans, walruses, and whales. Indigenous hunters along the Bering Strait transform the energy in the bodies of whales by eating them in order to survive, while foreign whalers from US and Soviet ports turn the embedded energy of Beringia into commodities that serve state capitalist and socialist systems instead of sustaining life.[44] "Energy" in Demuth's account is a way of approaching interrelated human and nonhuman histories across time, of finding a way to understand energy outside of fossil modernity.[45]

The metabolic or elemental sense of the word "energy" offers a way out of the anachronism problem for scholars who engage pre-fossil-modern histories. Exploring energy as a force produced both through the use of such natural resources as wood or oil as well as through biological processes of metabolism in plant, animal, and human bodies makes "energy" a category available for the study of history wherever there is metabolism—across human history and throughout the more-than-human world. Understanding energy and metabolism as processes of life, following Kyla Wazana Tompkins and Samantha Frost, invites us to see human and more-than-human bodies as processes in a constant state of becoming.[46] This elemental sense of the word "energy" is especially important with respect to histories of whaling, because as Demuth's work shows, it centers Indigenous practices of whaling for survival rather than commercial practices of whaling for profit. To approach energy as Demuth does is liberatory; her history of Beringia releases energy from its vicelike conflation with fossil fuels.

There are other ways to understand whaling as energy (or proto-energy) and that can help pry the concept of energy away from fossil fuels. Doing so also requires inquiring into why, where, and how the concept of energy has come to be linked so closely with fossil fuels. The history of whaling helps address this part of the energy problem as well. The main products of the whaling industry—whale oil and spermaceti—were part of a volatile and heterogenous marketplace for illuminants in the nineteenth century that also included beef and pork tallow, coal oil, natural gas, phosphorus, and petroleum.[47] The "illuminant" economic sector was a direct precursor to the "energy" sector, and its very heterogeneity helped give rise to the unifying concept of energy that collapsed vast differences in labor and extractive practices. These diverse illuminants and oils were brought together, not only in the abstract concept of the "market," but in cultural texts such as the *Vanity Fair* cartoon of the dancing whales that drew whale oil and rock oil together in a single image. I suggested above that this concept of energy—energy as a unified field of labor, production, and circulation based, not on the conditions of a specific resource's extraction, but on the consumption of its abstract product—is a myth. In fact, it is a dangerous myth about energy, a myth that palliates us into believing that all energy resources are ultimately the same. Or, at the very least, the myth makes it convenient to ignore the material origins and consequences of energy. To think of energy this way creates what Mark Simpson and Imre Szeman have called "energy impasse," a condition of being stuck in a fossil-fueled system that we know beyond any doubt is the source of climate change and extractive violence. Simpson and Szeman charge the very notion of energy transition with energy impasse:

> The stuckness to which we want to draw attention is conditioned by a narrative at the heart of transition's logic: that there have been *other* energy transitions, that the time of energy is *always* about transition. The dirty energy era began, the story goes, with an originary shift from wood to coal and, after successive periods of realignment in the dominant forms of energy used, will find its apotheosis in the transition from oil to clean energy. . . . Transition's fiction thereby makes sure that we remain stuck in a present that withholds, in the active language of its idiom, any capacity to create genuinely different energy future.[48]

I find it necessary, as I think back to the nineteenth century, to continue to use the term "transition" to describe the undeniable change in environment, culture, and climate that took place as fossil fuels began to be consumed at higher and higher rates. But I intend to use the word "transition" with a sense of history and a due sense of warning about the fictions it creates. In fact, it is my task here to locate, name, and describe the fictions of transition that concern Simpson and Szeman: the narratives that keep us stuck in fossil modernity. The whales in the oil fields, the dancing whales, and the many texts I discuss in this book: these are some of the sites where the narratives and fictions of energy—and energy impasse—are born. And so in this book, I treat whaling as a proto-energy industry because it was part of the imaginative project that joined different fuel sources and labor practices under a single unifying concept of energy. I write this book hoping for a future energy culture that undoes the harms of extractive capitalism and creates new forms of community organized around justice and care.[49] Imagining that future means reckoning with the forms of energy culture we need to unmake: that is the task to which I have set myself in *Rendered Obsolete*.

ENERGY ARCHAEOLOGY

There is another way out of the anachronism problem that attends the study of whaling within the rubric of energy, and that is to reframe energy as a method rather than as a subject. With this book I propose the method of *energy archaeology*: a way of telling the history of energy from the point of view of the present in order to locate the traces of old energy resources, technologies, and cultures in contemporary and emerging energy cultures.

In setting out the parameters of energy archaeology, I follow two main methodological models. The first is the model set forth by scholars of media and network archaeology.[50] Scholars in media studies have developed a robust set of theories for obsolete technologies, including the "residual."[51] A subfield of media studies, media archaeology approaches artifacts and practices that did not circulate under the name "media" in their time but that operated in their own times in a way that resembles "media" in ours or that presage later media in interesting ways. Media archaeology offers a special opportunity to uncover what Erkki Huhtamo has called "neglected, misrepresented, and/or suppressed aspects of both media's past(s) and their present."[52] For media archaeologists, who follow Michel Foucault's terms, "archaeology" and "excavation" are the metaphors that describe a historical methodology that begins from the present—the surface—and digs down without ever losing contact with its terms. Scholars working in and at the edges of media archaeology approach the tasks of excavating with differing degrees of literalness and at different scales. Media theorist Jussi Parikka, in *A Geology of Media*, goes literally to ground in order to excavate the geological basis of media.[53] This is much the sense in which historian Bob Johnson has described his own history of fossil fuels in US culture, *Mineral Rites*, as an "archaeology of the present."[54] I also find very generative media studies scholar Nicole Starosielski's proposal for "network archaeology" in her study of undersea fiber-optic cables, because this approach puts an emphasis on massive global infrastructures that are especially relevant to energy.[55]

I am not the first to bring media and energy studies together; the energy humanities have grown up in kinship with media studies, mainly through shared inquiries into infrastructure. In the broadest possible terms, media and infrastructure are both structures (physical or symbolic) that direct attention and arrange communication. Their definitions and objects of study diverge, but they share this common ground on which scholars have worked to understand these complex systems. In various contexts, scholars have argued that media would not exist without infrastructure; that media are infrastructures; and that infrastructures are media.[56] Writing about twentieth-century US petromodernity and oil culture, Stephanie LeMenager has theorized energy as media—"oil media"—following the capacious definition of media first set out by philosopher Marshall McLuhan: "Compelling oil media are everywhere." LeMenager writes, "Films, books, cars, foods, museums, even towns are oil media. The world itself writes oil, you and I write it."[57] The nineteenth century is a particularly exciting site for media, infrastructure, and energy studies, since "media," like "energy," underwent massive transformations in meaning during this time. Daggett traces the consolidation of energy's meaning in the nineteenth century around fossil fuels and work. Along similar lines, media historian John Durham Peters proposes an etymology for "media" by noting that in the nineteenth century the word "media" typically referred to natural elements such as water, earth, fire and air—an etymological fact that Peters invokes in claiming the "elemental legacy of the media concept."[58] The concepts of media and energy bear traces of their old unconsolidated meanings in the nineteenth century in ways I explore in this book.

Whales and whaling play an unacknowledged role in the formation of knowledge about media and infrastructure, as well as energy. In chapter 1, I read *Moby-Dick* as a work that should be understood (at least in part) as a theoretical text in energy and infrastructure studies, in its lucid representation of how raw energy resources and complex infrastructural systems constitute each other. In chapter 3, I interpret ships and even whales themselves as media that structure communication about the decline of organic energy regimes and the ascendance of fossil fuels. My work in chapter 5 also finds that Lewis Mumford, often cited as a progenitor of infrastructure, media, and energy studies, developed his understanding of these terms through his earlier study of Herman Melville and *Moby-Dick*.

But while my practice of energy archaeology is embedded in scholarly discussions around media and infrastructure, this book's most important methodological forerunners and interlocutors are its own subjects. Energy archaeology is a vernacular practice that emerged alongside the concept of energy itself in the early days of the fossil fuel energy regime that this book chronicles. Herman Melville, for example, was an energy archaeologist. Read in the light of energy archaeology, *Moby-Dick* is a work of nineteenth-century futurism that forecasts the inevitable ends of extractive energy regimes: the extinction of life and the obsolescence of culture. The local promoters who converted declining whaling ports such as Nantucket into coastal tourist attractions in the 1870s, and the tourists who visited them, were energy archaeologists who found in the crumbling infrastructures of an exhausted resource industry an analogue for leisure in an age of increasing mechanization. The sideshow hawkers who toured the body of a dead whale on a nationwide rail tour in the 1880s were energy archaeologists for whom the huge whale corpse was both metonymic of the dying whaling industry and an absurd limit-test for the mighty coal-powered railroad. To claim that this book's subjects are my own methodological models is not to claim that they are all good, ethical energy archaeologists. In chapter 4, for example, I show how whaling history was coopted in the early twentieth century to cement an enduring link between white supremacy and energy extraction.

Like the discipline for which it is named, energy archaeology is not a value-neutral scholarly practice, but neither is it doomed to follow all of the ethical ills of resource extraction. Environmental humanists and activists have named and begun to organize around the "slow violence," as Rob Nixon coined it, produced by energy extraction and consumption: the toxicity and pollution leaching out into the air and water around sites of extraction, the violent effects of climate change itself driven in part by fossil fuels, the intensification of colonial violence at sites of fossil fuel extraction and infrastructure development.[59] Energy archaeology joins these efforts to name, expose, and counter the violence of energy extraction by exploring the cultural and ideological mechanisms that underpin it.

I seek to practice a form of critical energy archaeology by learning from and engaging work by Indigenous scholars, activists, and writers. Throughout the book (especially in chapters 1 and 4), I draw on Indigenous scholarship that offers the most incisive analysis of how settler colonialism and extractive capitalism are entangled in climate crisis.[60] Indigenous scholars, writers, and activists have also located the extractive violence of infrastructure development and even of historical writing and historic preservation, as illustrated in Jean O'Brien's extraordinary *Firsting and Lasting*.[61] And I am inspired by Métis scholar Zoe Todd's proposal to extend the care and responsibility for more-than-human beings to the "fossil kin" that have been weaponized through fossil modernity.[62] Todd's proposal for "fossil kin" opens up the radical new possibility that energy can be theorized from a position of care instead of power. This work helps me locate the violence of settler colonialism at work in energy culture and to see fossil modernity itself as a variant of what Mark Rifkin has called "settler time."[63]

By tracing the present crises through longer structures of colonial extraction, energy archaeology offers a new, historical approach to the energy humanities.[64] Many of the projects in the field as it is currently constituted explore fossil fuels and energy culture in the twentieth and the twenty-first centuries—driven, rightly, by the political urgency of climate change, itself a product in large part of fossil fuel consumption. But "energy" is a category with a much deeper history than the technocratic term would suggest. Energy archaeology suggests one approach to that history that still honors the environmental crises and political urgencies of the twenty-first century. Energy archaeology also offers one way of addressing the problem that Christopher Jones has called "petromyopia," the tendency of the field of energy humanities to focus on fossil fuels and, in particular, on petroleum.[65] Energy archaeology offers a way out of petromyopia: it is a method that unlocks the study of fuel sources and systems of work before—and, crucially, after—the era of massive fossil fuel consumption. It is useful for our scholarship and perhaps even for environmentalist activism to unhitch our conception of energy from fossil fuels as we confront the cascading crises of extractive capitalism. Energy archaeology offers one way forward.

OCEANIC ENERGY

The story of US whaling in fossil modernity is, of course, a story about the sea—but in a sense, most stories about energy are. The ocean exerts a tidal pull on energy studies. Fossil energy revolutionized oceanic shipping—through the transformation of maritime technology from sail to coal-powered steam to petroleum-based fuel—and so, too, did the ocean change oil. The Royal Navy's decision to convert Britain's fleets from coal to oil power in World War I has been credited by energy scholars as a signal moment in the history of oil's ascendancy as the primary source of fuel and of energy security as an existential concern for nations and militaries.[66] Fossil modernity happens at sea: the ocean's surface is latticed by the trajectories of oil and gas tankers and dotted with oceanic oil drilling platforms, and its depths are stabbed by ultradeep oil wells and pipelines. The airplane might signify the apex of fossil modernity's obsession with speed and efficiency, but the vast majority of commerce still travels by sea in container ships and in no small part because the dirty fuel burned by commercial ships is so much cheaper than jet fuel. Fossil modernity is the story of ever bigger ships and ever smaller boats: of massive post-Panamax or post-Suezmax container ships, so named because they are too wide to fit through the world's most significant canals, and of tiny boats carrying people driven to dangerous oceanic migrations by social upheaval and climate change.

In the past half-century or so, at least since the Santa Barbara oil spill of 1969, the ocean has also been figured in environmentalist discourse as a site of fossil modernity's violence and damage.[67] The ocean is the place where oil spills, and its beaches are the places where oil-soaked sea birds wash up on greasy sand. Climate violence makes the ocean ever more dangerous for creatures who live in and near the sea. The ocean is polluted by oil and choked with plastic garbage, and meanwhile the acidified ocean poisons its own marine life and the rising sea threatens coastal and archipelagic communities.

But the impacts of fossil modernity on the sea are often hidden from sight, invisible to those who do not labor on ships or at ports or coastal refineries or experience the direct violence of rising tides or hurricanes turbocharged by climate change.[68] Oil companies work hard to hide breakdowns such as oil spills. BP, for example, established a military-style perimeter around the site of the Deepwater Horizon blowout in 2010 in order to shield hundreds of miles of oily water and coastline from public sight or journalistic documentation. As Anne McClintock writes, "With the conjoined collusion of BP, the US Coastguard, the National Guard, and the Obama Administration, the Gulf disaster fell into a great, administered forgetting."[69]

Shipping containerization is a powerful consequence and intensifier of fossil modernity, and shipping containers have rendered oceanic shipping and labor more invisible, too. Shipping containers were invented by trucking executive Malcolm McLean in 1956 to make it easier to get goods from ships onto trucks or trains: to smooth out the bump in the supply chain where the sea met land.[70] In the hybrid photo-essay *Fish Story* (1995), the photographer and essayist Allan Sekula chronicled changes that the process of containerization wrought on the shipping industry in the second half of the twentieth century. Shipping containers anesthetized harbor spaces by literally hiding cargo from sight (and, as Sekula notes, smell), and they depopulated those spaces, too. Because automated cranes could lift shipping containers faster and more efficiently than human stevedores, containerization radically reduced the number of laborers needed to load and unload cargo from ships. Containerization rewrote the relationship between harbors and cities, too; whereas harbors had once been at the center of coastal cities—and even formed the main entrance of ocean-facing communities—containerization put pressure on developers to move ports out of town so that container ships were as close as possible to overland mass transport hubs such as highways and freight railways. Containerization hid cargo in boxes and sequestered busy ports far out of town: part of the process that Sekula calls the "'forgetting' of the sea."[71] "Forgetting" the sea is a privilege unavailable to laborers on ships and alongshore who depend on the sea for their living, but today there are undeniably fewer people around to do the work of remembering; less than 0.5 percent of the world's population handles 90 percent of the world's overseas freight.[72]

That the sea has become hidden, or forgotten, or ignored in scholarship, popular culture, and public consciousness has by now become a conventional idea, as well as the basis for new oceanic scholarship in the humanities. In her study of *The Novel and the Sea*, professor of comparative literature Margaret Cohen has elaborated on Sekula's "forgetting the sea" concept by naming "hydrophasia" as a pathology that has prevented literary critics from paying attention to the oceanic elements of fiction.[73] The oceanic gap in humanities scholarship prompted literature scholar Hester Blum to call for a new field of environmental humanities research called "oceanic studies," which should be "a practice . . . that is attentive to the material conditions and praxis of the maritime world, one that draws from the epistemological structures provided by the lives and writings of those for whom the sea was simultaneously workplace, home, passage, penitentiary, and promise."[74] The field of oceanic studies has flourished over the past decade and has given rise to important new work in history, literary studies, Indigenous studies, critical race studies, and media studies that fulfills Blum's call for oceanic studies.[75] Some of the most important new works in oceanic studies come from Black feminist and queer scholars who resist the "forgetting" narrative by revealing the centrality of the ocean to Black thought through the memory of the Middle Passage. Christina Sharpe deploys the multivalent metaphor of "the wake" (only one of whose meanings concerns "the track of water left on the water's surface by a ship") to describe the ongoingness of the structures of transatlantic slavery in Black lives and thought.[76] The sea cannot be forgotten for those who live in the wake.

Although "forgetting of the sea" has been a powerful impetus to new art and scholarship, "forgetting" is not quite the right word for what happens to the sea in the fossil-modern cultures I describe here. Literature, art, popular culture, tourism: at these sites of cultural production, the sea is hypervisible. Consider the ubiquity of the beach or cruise vacation in the middle-class US cultural imaginary, and the whole fleet of cultural products that support the beachy dream, from beach reads in publishing to resort wear in fashion, from yacht rock and Jimmy Buffet to Corona beer ads imploring viewers to "find your beach." The beachy dream is available in an infinite variety of forms for an infinitely segmented consumer base: the tropical sea of air-brushed romance, the off-the-grid beach hut that restores burned-out minds, the preppy sea of Kennedys and yachts, the kindergarten-classroom sea of sandcastles and sea creatures. These beachy dreams may not ostensibly be concerned with containerization, sea level rise, or other twenty-first-century oceanic problems, but these fantasies, too, underpin fossil modernity.

This book offers a different angle on the story of the "forgetting of the sea," drawing on a problem with the idea that Sekula himself acknowledges. Oceans and coastlines have not been forgotten; instead, the sea has become a place that signifies the past. Sekula notes: "Obviously, the sea does not simply disappear over the course of the twentieth century. The broader culture of modernity, modernist or not, sustains a repetitive fascination with maritime themes. If many of the features of the pre-modern and romantic attitudes toward the sea are no longer credible, they surface nonetheless, as if the sea were indeed a bottomless reservoir of well-preserved anachronisms."[77]

The history of nineteenth-century energy and, indeed, the history of US commercial whaling and its afterlife, show how the sea became "a bottomless reservoir of well-preserved anachronisms." This book offers an expansive understanding of the maritime world by seeing it as a place of leisure as well as labor. The changes in energy culture of the nineteenth century mean that whaling ports once deemed productive had to be reinterpreted within the new regime of fossil energy: as sites of rest, leisure, entertainment, and play.

By showing how oceangoing whaling vessels and New England working harbors were transformed into tourist sites, *Rendered Obsolete* identifies the whaling industry as a surprising point of continuity between the oceanic workplace and the oceanic vacationland. This book also shows how oceanic objects, infrastructures, and even epistemologies made their way far inland beyond the coasts, making the case for a much more expansive understanding of the "maritime world," one that encompasses Nantucket and Chicago, the ship and the prairie. In its later chapters, this book shows, too, how the history and heritage of whaling and the sea become contested sites of ideology and identity, available to a wide range of political beliefs and signifying the many interpretations and uses of the past.

CHAPTER OUTLINE

This book draws together a wide and unruly archive of cultural artifacts that circulated throughout the US commercial whaling industry's peak production and especially in its decline and obsolescence, from 1851 through 1930: personal narratives, novels, engravings, newspaper accounts, tourist paraphernalia, public performances, films, exhibitions, wooden whaling ships, and even whale carcasses and body parts. These sources documented, dramatized, and, in some cases, romanticized the whaling industry even as its main commodity was rapidly supplanted by petroleum. The very existence of this archive proves that whaling did not simply vanish in the magic trick of creative destruction. Nor did whaling culture dematerialize only to reemerge as traces in evanescent concepts of energy. Whaling culture is material as well as textual: infrastructural, mobile, durable, dangerous, and sometimes even putrescent. But the texts under discussion here did more than simply exist; they were some of the material and conceptual conditions that gave rise to fossil modernity.

The story of US whaling culture during the industry's obsolescence is suggestive of the course of other obsolescing technologies. Whaling is, I argue, particularly resonant as a case study in the cultural side of energy transitions. I hope it may serve as a provocation and warning for readers encountering this book at the beginning of what we can only hope is another paradigmatic energy transition: the infrastructures, conceptual templates, cultural forms, and attachments and relationships that arose in fossil modernity will not simply disappear as the world slowly relinquishes fossil fuels. What will the cultural afterlife of petroleum bring? The culture of obsolescing energy regimes can exert surprising power on ascendent ones. But the story that this book tells is very particular, of how, where, and why whaling culture circulates throughout the United States in the late nineteenth and early twentieth centuries, and how that culture changes and is changed by encroaching fossil modernity.

The story arrives in three parts: chapter 1 serves as a theoretical preface in which I present Herman Melville as an energy archaeologist and read his *Moby-Dick* (1851) as a work of energy theory. Written by Melville at peak whale oil production, *Moby-Dick* presents the commercial US whaling industry as an energy industry *avant la lettre* but in terms that anticipate and even surpass the efforts of twenty-first-century energy theorists to understand the reach of energy resources, infrastructures, and financial forms into every part of ordinary human life. As a chronicle of the US commercial whaling industry, Melville's novel is hyperbolic; whaling was a powerful industry but nothing like coal or oil would become in the fifty years following the book's publication. But it is a work that chronicles and critiques energy culture at the cusp of fossil modernity. Specifically, *Moby-Dick* critiques the mandate for growth in extractive capitalism, and in this chapter I contextualize the novel within a history of ideas about economic growth and degrowth. Returning again and again to the looming threat of resource exhaustion, the novel meditates on the exhaustion and obsolescence that come for all worlds built on extraction. As a forecast of fossil modernity, I accord this chapter a Melvillian place in the structure of this book: "Loomings."

The second part of the book describes the emergence of what I call "Whaling Entertainment." If, as Cara Daggett has claimed, the ruling concept of energy was sutured together with work at the beginning of fossil modernity, whaling culture served as fossil modernity's foil: a yardstick to measure the distance US culture had come since the old, inefficient days of US whaling. At the same time, this archaic form of work became a very modern form of entertainment. Whaling entertainment reached wide public audiences in the late nineteenth century: whaling culture traveled away from the East Coast toward the expanding US inland territories and, at the same time, drew new audiences to the eastern home ports of the fading whaling industry. The movement of whaling culture in both directions—away from and toward the fading industrial centers of the US commercial whaling industry—was made possible by the expansion of, and increasing access to, coal-powered railroads and passenger steamers, the energy-transportation assemblage that pushed the whaling industry into decline. Chapter 2 shows how poverty, depopulation, and infrastructural ruin on the declining whaling island of Nantucket came to appear as "quaintness" to a new class of wealthy and middle-class tourists. It served the Nantucket tourism industry to portray the island and the sea as what Sekula has called a "bottomless reservoir of well-preserved anachronisms": quaintness and anachronism were the basis of Nantucket's appeal to worn-out city workers who, following the emerging logic of energy, perused Nantucket's whaling history as part of a restorative vacation that "recharged their batteries" for productive work back at home.

Chapter 3 tells the story of two spectacular traveling exhibitions that brought whaling culture to the Midwest—in particular, to the ascendant city of nineteenth-century US industry: Chicago. The first was the rail tour of a giant (and dubiously preserved) whale's corpse. A speculator named George Newton partnered with a midwestern promoter to form the Pioneer Inland Whaling Association, whose purpose was to tour the body of a dead whale around the country. The dead whale, also known as the Prince of Whales, toured the country for two years in various states of decay and remediation before its proprietors sold the interest and gave it up. The whale show is exceptional in a sense; it brought an oceanic creature a thousand miles inland. But in another sense, the whale show is perfectly continuous with the logics of animal extraction and serves as the logical endpoint of rendering. In the form of oil, baleen, ambergris, and scrimshawed bones and teeth, whale bodies had been traveling great distances inland from the ocean for centuries. The combined forces that rendered the whaling industry obsolete—fossil fuel extraction and railroad expansion—made the Prince of Whales tour possible and, at the same time, represented a new phase in the rendering of the whale's body for public consumption. In the whale show, the whale shifted ontological categories; the show rendered the whale, not for its body's constituent substances, but according to the other definition of the term, by making it into a representation of the real thing. One decade after the Prince of Whales debuted in Chicago, another unlikely element of whaling culture made the long inland voyage: a New Bedford whaling ship, which joined the World's Columbian Exposition in 1893 as an exhibition. Alongside the dynamos and electrical lights that ushered in new American technological futures, the whaling ship—perversely called the *Progress*—benchmarked technological antiquity, offering in material form a reminder of the difference between the quaint brutality of whaling and the ostensibly cleaner, brighter fossil resources that would power America's future. The stories and even the names of these Chicago whaling exhibitions—*Progress* and Pioneer Inland Whaling Association—are whimsical, but at the same time these names link the discourse of oceanic resource extraction to practices of frontier settler colonialism so prominently on display at Chicago.

If whaling came to Chicago to highlight progress, back in New England it was commemorated through acts and rituals of what I call "Whaling Nostalgia." Nostalgia is the focus of this book's third section. Chapter 4 tells the story of how whaling culture was commemorated in the former center of the US industry, New Bedford, Massachusetts: whaling culture was enshrined in public monuments, museums, historical writing, and even a Hollywood film funded and produced by New Bedford city leaders. My study focuses on the racial politics of these commemorations. These portrayals recast US commercial whaling in a heroic mode: historic whalemen remade into avatars of idealized white masculinity and the whaling industry itself credited with pioneering US overseas imperialism. The story of US commercial whaling's decline—so apparent in New Bedford, where the last few whaling voyages were departing in the 1920s—intersected with a cultural narrative of white aggrievement that fueled white supremacy and anti-immigrant nativism in early twentieth-century US politics. For preservationists motivated by white supremacy, whaling became in these commemorative projects another site of loss, degraded from within by nonwhite immigrant laborers and from without by the inhuman forces of mechanized modernity. The commemorators of New England whaling often cast the decline of the industry in ecological terms, albeit not in the ecological terms one might expect: the decline of New England whaling was a problem of the extinction, not of whales, but of "Yankee" whalemen and the values they represented. Of course, this extinction narrative was entirely specious; civic monuments to the US whaling industry were funded and promoted by powerful white men who had built their wealth and political power in the corporate consolidation and financial opportunities of early fossil modernity. At the moment of the industry's expiration in the economy, these preservationists claimed whaling heritage for whiteness. Whaling became a way for white men of the Northeast to articulate their authenticity, to control the infrastructure and finances of fossil modernity, while at the same time staking a claim to pre–fossil fuel history; whaling became part of the cultural mechanisms that linked extractive capitalism with white supremacy. Nostalgia for the lost days of "Yankee" whaling licensed violent expressions of white supremacy and misogyny. These ideas are usually associated with the myth of the Lost Cause and the formation of southern identity at the turn of the twentieth century, but the commemoration of the whaling industry helps show how white supremacy took root in New England regional identity, too, in part through the nostalgic representation of whaling culture in the early twentieth century. The case of whaling and New England regional identity in the early twentieth century shows how energy transitions can stir up violent expressions of white supremacy. White aggrievement, nostalgia, and narratives of loss are no less powerful today than they were in the early twentieth century, and energy transitions can turn into powerful stories of loss in the wrong hands. The coal nostalgia at the heart of contemporary American politics and energy policy is a direct descendent of the extractivist nostalgia in early twentieth-century whaling culture and a warning about other energy transitions to come.

While the fourth chapter deals with the commemoration of New Bedford as the site of the whaling industry's death, the final chapter turns to the recuperation of Melville—himself having been rendered obsolete by the nineteenth century literary establishment—in what is known as the Melville Revival of the 1920s. *Rendered Obsolete* is bookended with chapters about Herman Melville's *Moby-Dick*. Chapter 5 of this book offers a second look at *Moby-Dick* and the works of scholars in the so-called Melville Revival. It is well known that Herman Melville and *Moby-Dick* underwent a dramatic critical reassessment in the 1910s and 1920s. Melville had died nearly in obscurity, and *Moby-Dick* had been considered one of his lesser books and a work suited at best for boys interested in the sea. The book was taken up and classed a "masterpiece" by critics building the new institutions of American literary studies.[78] I argue that the works of the Melville Revival are also poignant works of energy archaeology, reading *Moby-Dick* for its insights into energy transition and social change, albeit different from the ones I identify in chapter 1. In 1851, when the novel was published, *Moby-Dick* embedded a sharp critique of natural resource extraction within an apocalyptic vision of the future: of what the world might look like after the ultimate extinction of whales and even humans and, on a more local scale, of what the thriving industrial whaling ports of the United States would look like when those resources and the wealth that they generated had moved on. By the 1930s, *Moby-Dick*'s vision of the future had in some ways come to pass; the whaling economy had suffered something like a regional extinction in New England port towns. The commercial whaling industry declined in the way that all extractive industries eventually decline: through the substitution of a newer, cheaper commodity for the commodity at the center of the whaling industry and, to some extent, through the exhaustion of the resource. And so, to many readers, *Moby-Dick* obsolesced and became quaint along with the industry itself. The book's complex dynamics of work, leisure, and natural resource extraction were reworked in accordance with the new place of whaling in the US economy.

These issues crystallized in the Lakeside Press edition of *Moby-Dick*, designed and illustrated by the artist Rockwell Kent, which I argue should be understood as a key text of the Melville Revival. I surface Kent's interpretation of Melville's novel and contextualize it within the work of the Melville Revival by approaching the form, content, and method of producing Kent's illustrations for *Moby-Dick*. I read Kent's illustrations as "skeuomorphs": artifacts made in one medium meant to emulate the formal features of another. Kent's illustrations closely resemble nineteenth-century woodcut engravings, but they are in fact ink drawings reproduced and published with then state-of-the-art imaging and printing technologies. There is a great difference between the artisanal woodcut engravings that Kent's illustrations appear to be and the highly mediated skeuomorphic images that they actually are; that difference signals a disjunction between Kent's nostalgic idea of honest work and the realities of work in fossil modernity. Kent's turn to *Moby-Dick*, and to the world of maritime labor more generally, indicates the dynamic and fraught status of maritime trade, coastal tourism, and extractive industry in this period. Kent's illustrations for *Moby-Dick* created a visual language of nostalgia for the obsolete wooden infrastructures of whaling in a form that resembles woodwork but is, in fact, all style. The drawings mark Kent's abandonment of a certain kind of artistic labor and, at the same time, the persistent pull of nostalgia for the outmoded labor of both woodcutting and whaling.

Rendered Obsolete ends with a twenty-first-century epilogue about the enduring and ever-changing cultural career of the last extant US whaling ship from the nineteenth century, the *Charles W. Morgan*. The ship was built in 1841 and currently resides in the collection of the Mystic Seaport Museum, in Mystic, Connecticut. The museum director decided to restore the *Morgan* during the economic crisis of 2008–9 because he hoped that such an audacious project—a "moonshot"—might bring needed attention and funding to the struggling historical museum. In 2014, the newly restored ship sailed throughout New England on an exhibition tour. In the twenty-first century, Mystic Seaport has used this traveling ship exhibition to tell stories about marine conservation, the save-the-whales movement, the everyday lives of US whaling laborers, ongoing practices of Indigenous whaling, and contemporary economic crises that make museums precarious. Greeting the ship at a dockside ceremony in New Bedford, Senator Elizabeth Warren cited the prosperous nineteenth-century whale oil industry as a precursor to the state's twenty-first-century sustainable offshore wind energy initiative. I myself sailed on the ship, and the experience helped me better understand my own entanglements in modern energy. The *Morgan* and its 38th Voyage offers me an opportunity to reflect on energy infrastructures past and future: on the potential power of preserving or, as energy activist and fellow energy humanities scholar Jeffrey Insko has proposed, dismantling obsolescing energy infrastructures.[79] Even the rich wooden housing of the museum ship *Charles W. Morgan*, suffused with the evocative aroma of pine tar, does not offer a space of retreat from fossil fuels and the crisis of climate change. Even at the end of fossil modernity, whaling is energy.

PART ONE
Loomings

Built-In Obsolescence
Energy and Limits to Growth in the
Whaling World of *Moby-Dick*

Moby-Dick's narrator Ishmael first encounters whale oil, the commodity at the center of the whaling industry, on the wharves at New Bedford:

> Huge hills and mountains of casks on casks were piled upon her wharves, and side by side the world-wandering whale ships lay silent and safely moored at last; while from others came a sound of carpenters and coopers, with blended noises of fires and forges to melt the pitch, all betokening that new cruises were on the start; that one most perilous and long voyage ended, only begins a second; and a second ended only begins a third, and so on, for ever and for aye. Such is the endlessness, yea, the intolerableness of all earthly effort.[1]

The whale oil casks are heaped in piles as large as "hills and mountains" that overwhelm the wharves. Ishmael's attention drifts from oil casks to the docked whaling ships, and from there, his imagination leads him beyond the town to the world's oceans. The whaling industry seems to Ishmael, as he ends the passage, endless. "Endlessness" is the industry's imperative and mandate: to increase productivity and grow, as Ishmael says, "for ever and for aye."

This short passage evokes the history of an economic idea: the idea that economic growth is vital to success and even well-being in capitalist economies. At least since Karl Marx, economists have believed that capitalist economies must expand in order to function and even to survive. Marx traced the processes of accumulation that ultimately lead to the growth of capital, and economists in the twentieth century later quantified capital accumulation by measuring gross domestic product (GDP), which, according to the growth paradigm, grows when nations are flourishing.[2] The whale oil passage from *Moby-Dick* asserts the growth paradigm through the word "endless," and it even accommodates readers to the mandate of endless growth with palliating alliterations: "side by side," "world-wandering ships," "silent and safely," "carpenters and coopers," "fires and forges." The repeating consonants sonically emulate the port's unceasing rhythms of dispatch and return, dispatch and return.

But the passage ends on a minor key, Ishmael revising his description of the "endlessness" of whaling activity by proclaiming "the intolerableness of all earthly effort." Endless growth is intolerable. Ishmael first asserts the growth paradigm and, with a swift revision, critiques it. The idea that growth has limits—that growth is not the best way of measuring individual or collective flourishing—is an old idea, too. The critique of growth arguably began with Thomas Malthus in 1798 and returned in often neo-Malthusian works on environmental limits in the 1960s and 1970s. The call for degrowth is growing louder in the twenty-first century in the face of climate change and its threats to the smooth operations of global capitalism.[3]

It is telling that a mountain of oil casks incites Ishmael's meditation on growth and its limits, because energy has always been tied up with the problem of growth. Growth needs fuel. Economic growth, especially as measured by increasing GDP in the mid-twentieth century, is literally fueled by increasing fossil fuel consumption and premised on unlimited access to fossil fuels.[4] Timothy Mitchell has argued that the dependence of GDP growth on access to energy characterized the postwar United States when "the decreasing cost of energy, in particular of oil from the Middle East, helped generate the effect of the economy as an object capable of unlimited growth."[5] But this kind of growth—tethered to the extraction of a mineral resource that is at the same time finite and also responsible for catastrophic climate change—has a dim future. The oil shocks of the 1970s and, even more recently, the mounting costs and cascading crises of climate change are revealing the concept of endless economic growth to be a fossil-fueled fantasy. Actors on all sides of the political spectrum seem to agree on the link between energy and growth; fossil fuel boosters cite growth as a reason to "Drill, baby, drill!" while progressives and climate activists advocate for fossil fuel divestment and economic degrowth. Climate change activist Greta Thunberg in her address to the United Nations Climate Summit in 2019 chastised world leaders still clinging to the growth paradigm of economic development: "People are suffering. People are dying. Entire ecosystems are collapsing. We are in the beginning of a mass extinction, and all you can talk about is money and fairy tales of eternal economic growth. How dare you!"[6]

Few have looked to Herman Melville as a voice in the history of energy or the growth paradigm, but I argue that *Moby-Dick* is a work that belongs in this genealogy. Specifically, it is an important work of energy archaeology: a work that interrogates the cultural, political, economic, and environmental dimensions of energy resource regimes, even as the very concept of "energy" itself is still consolidating.[7] More than a century before the growth debates, Melville made the connection between the oil resource and the fantasy of endless growth. Ishmael's first fleeting encounter with the material commodity of the whaling industry—whale oil—incites in him a profound reflection on economic growth and its limits. In many ways, the rest of the novel is an extended meditation on these themes. As Ishmael voyages on the *Pequod* and immerses himself in the workings of US commercial whaling, his insights about the proto-energy industry of whaling widen and deepen. Ishmael reveals the fatal fallacies of extractive capitalism by testing the landsman's view of whaling against the direct experience of the industry he acquires as a laborer on the *Pequod*. His conclusion: endlessness is the mandate of capitalism, but intolerableness is its reality.

Yet *Moby-Dick* is more than a work that fits neatly into a stable historiography of energy history and growth critique: the novel shows how energy and growth became linked in the nineteenth-century proto-energy industry, and it explains how and why the growth paradigm will ultimately fail. *Moby-Dick* gives voice to and explores the cultural forces that compel the whaling industry: infrastructure development, settler colonialism, economic growth. And at the same time, the novel presents the specter of exhaustion and extinction that haunts the industry, and it cycles through several scenarios for whaling's end. *Moby-Dick* is perhaps underexplored as a work of energy theory because its central resource seems marginal or inconsequential in comparison to fossil fuel industries such as oil and coal—and perhaps, too, because the narrative voice of *Moby-Dick* is so ambiguous. But it is precisely the rapid combination of assertion and critique on display in this passage and elsewhere throughout the novel—for example, "endlessness" and "intolerableness"—that makes *Moby-Dick* a profound work of energy archaeology. Through the greenhorn whaleman Ishmael, Melville perceives the disjunction between the smooth flows of capital on land and the violence of extraction at sea. He apprehends the commercial whaling industry as a powerful resource-infrastructure assemblage literally remaking the American landscape through the whale's own body. And, understanding that material resources are always limited, he seeks to imagine a postwhaling future and ruminates on what will happen to whaling laborers and infrastructures and whales once the industry has ended. The narrative of the *Pequod*'s voyage offers up a magnificat on energy and the limits to growth.

Moby-Dick shows, too, that the US whaling industry of the nineteenth century is an important site for theorizing extractive capitalism, even if US whaling was not as disastrous to our present world as the global petroleum industry, and even if nineteenth-century whaling did not slaughter nearly as many whales as the whaling industries of the twentieth century.[8] The scale of destruction is important, but it is possible to observe the violence of extraction from the point of view of any extractive industry. *Moby-Dick* shows us that to think about extractive capitalism from the perspective of whaling and whale oil helps us (1) to visualize the violence of extraction by witnessing the excruciating scenes of death (human and nonhuman) at sea; (2) to understand the way that extraction fuels settler colonialism by showing how the whale takes its place in the history of imperial infrastructure development and territorial expansion; and (3) to grapple with the inevitable limits of the resource, to name and confront its scarcity, and to imagine in vivid detail the future after the exhaustion of the resource in question.

Because *Moby-Dick* was written at the very cusp of fossil modernity, it is tempting to say that the book is prophetic in its critique and its description of the violence inherent in extractive capitalism. Indeed, there is even something of a tradition in Melville criticism of attributing "prescience" or "premonition" to *Moby-Dick*.[9] But even had the world taken a different course and petroleum not ascended to its place of dominance in global technologies of mobility and commodity production, the violence of extraction was already underway on small and large scales. *Moby-Dick* may see energy's future, but only because the future will resemble its present and past. There is no better term to describe *Moby-Dick*'s orientation to the temporality of energy and the future of fossil capital than Melville's own: not "prophecy" but "loomings." "Loomings" are objects glimpsed at or even, in mirage, beyond the horizon line. Although the appearance of objects through looming seems magical, loomings are optical illusions produced through perfectly explicable effects of atmosphere and light. So, too, the effects of extractive capitalism and its fatal limits: they were objects readily observable to Melville in his world. From the historical situation of *Moby-Dick* and the mid-nineteenth-century commercial whaling industry, fossil modernity looms over the horizon.

This chapter follows the structure of the passage with which it began— first by exploring endless growth in extractive capitalism, and then by ruminating on its intolerableness. I proceed first to explore the way that

Moby-Dick represents growth. Ishmael shows what economic growth looks like in the form of whale oil wealth in New Bedford. Later, from within the process of whale butchering itself, Ishmael imagines the relationship between the slaughtered whale and the growing infrastructure networks on land that support urban development and colonial settlement. I then explore the limits on growth in the whaling industry: the approaching depletion of whales and the exhaustion of the whale oil resource, along with the looming "creative destruction" of capitalism itself, which consigns all booming industries, innovative technologies, and expert labor to the slow death of obsolescence. Most recent theories of obsolescence have emerged from the study of digital culture in new media, where the mandate for innovation resembles the larger capitalist mandate for growth (and where the innovation/obsolescence cycle has resulted in a catastrophic pileup of toxic e-waste). It is worth recalling, though, that planned obsolescence was also a feature of automobile design starting in the 1920s and arguably shaped eighteenth-century ship design, too.[10] Melville sets *Moby-Dick* in an obsolescing and in some ways already obsolescent world and, in so doing, apprehends its absurdity and inherent violence.

WHALE OIL MAGIC IN NEW BEDFORD

Before Ishmael sees those oil casks on New Bedford's wharves, he encounters the wealth that whaling has produced for its shipowners. New Bedford, to Ishmael, is a staggeringly wealthy place whose gaudy riches stand in stark contrast to the barren landscape of the town. It is clear from Ishmael's first glance that the town's wealth is far-fetched, because the land itself is poor in natural resources: "The streets do not run with milk; nor in the spring-time do they pave them with fresh eggs. Yet, in spite of this, nowhere in all America will you find more patrician-like houses, parks and gardens more opulent, than in New Bedford. Whence came they? how planted on this once scraggy scoria of a country?"[11] The wealth of the town, growing out of a "scraggy scoria," is almost magical, as if it had appeared from nothing.

The appearance of wealth as if by magic is a feature of extractive capitalism and the stories told about it. Jennifer Wenzel locates such forms of magic at work in literature from twentieth-century Nigeria, which, like New Bedford in the 1850s, is the site of an oil boom. Drawing on terminology from anthropologist Fernando Coronil, Wenzel examines the way that Nigerian literature engages and critiques the pervasive illusion of "petro-magic": "petroleum's false promise of wealth without work."[12] A similar illusion lies at the heart of Ishmael's perception of New Bedford, a place whose wealth was generated, like Nigeria's, by extractive industry. When Ishmael sees the grand patrician houses of the town, he does not imagine the many steps that it takes to translate a whale into a mansion: killing and butchering the whale; rendering, refining, casking, transporting, and selling its oil; buying land and housing materials, contracting labor, and building the house. Rather, he imagines the process of building New Bedford as something like a single marvelous feat: "Yes; all these brave houses and flowery gardens came from the Atlantic, Pacific, and Indian oceans. One and all, they were harpooned and dragged up hither from the bottom of the sea."[13] The whale and the material trappings of whale-oil-wealth—the mansions and gardens—are collapsed through metonymy into one body, captured and killed and brought from the sea onto land. Like the fantasy of petroleum, the fantasy of whale-oil-magic is one of wealth without work—patrician mansions and flowering gardens dragged up from the sea all in one piece, as if by a giant or sorcerer.

According to Wenzel, Nigerian literature critiques and "pierces [the] illusions" of petro-magic, not by denying its power, but by representing it through magical scenarios and characters.[14] According to Wenzel, Nigerian literature responds to the fantasy of petro-magic with "petro-magic-realism."[15] *Moby-Dick* does not puncture the illusion of whale-oil-magic in precisely the same way as petro-magic-realist Nigerian literature. Instead, Ishmael experiences, in rapid succession, enchantment and disenchantment. It is surprising that Ishmael erases extractive labor in the figure of magical mansions, because he magnifies the minutest details of whaling labor throughout the rest of the massive novel. *Moby-Dick* is, in some ways, an epic and explicit rebuff of the illusion of whale-oil-magic, a stiff disillusionment of Ishmael's early, naive impression of New Bedford's flowery wealth. Later in the novel, Ishmael stringently denies whale-oil-magic by exhaustively (and exhaustingly) cataloging the many perils of the whaling industry. He exhorts the intense human cost of whaling labor: "For God's sake, be economical with your lamps and candles! Not a gallon you burn but at least one drop of man's blood was spilled for it."[16] The cenotaphs on the wall of the Whaleman's Chapel—and the grieving widows who come there alone—tally the human cost of whaling in New Bedford, but discreetly, in an interior space dedicated specifically to whalemen and their widows rather than on placards in a flowery public square. The cenotaphs are wealthy New Bedford's repressed subconscious.[17] Whale-oil-magic ceases to cast spells on anyone who leaves New Bedford in a whaling ship.

Ishmael's representation of the industry's brutal violence on nonhumans offers another rebuttal of whale-oil-magic. For example, Ishmael narrates with pathos the murder of one giant, old, single-finned whale: "For all his old age, and his one arm, and his blind eyes, he must die the death and be murdered, in order to light the gay bridals and other merry-makings of men, and also to illuminate the solemn churches that preach unconditional inoffensiveness by all to all."[18] This blood-for-oil perspective on the whaling industry—a view that encompasses violence to humans and nonhumans—punctures the illusion of whale-oil-magic. This is the whaleman's view. Back on shore, in wealthy New Bedford, landsmen can comfortably ignore the costs and perils of whaling labor and sustain the illusion of whale-oil-magic.

Along with the conspicuous wealth of New Bedford mansions, weddings serve as Ishmael's shorthand for the wealth that whaling generates and the displays of conspicuous consumption of which whale oil is a part. In the passage above, Ishmael evokes brilliant weddings lit by whale oil, and in the early passage about the wealth of New Bedford, he chronicles another metonymic substitution of whales for wealth, like the conflation of mansions and whales: "In New Bedford, fathers, they say, give whales for dowers to their daughters, and portion off their nieces with a few porpoises a-piece. You must go to New Bedford to see a brilliant wedding; for, they say, they have reservoirs of oil in every house, and every night recklessly burn their lengths in spermaceti candles."[19] Whales in this passage appear both as symbolic currencies—whales and porpoises as dowries—and in their rendered form as the material wealth itself in the form of abundant candles and oil for lamps. According to Marx, commodities embody both use value and exchange value, and New Bedford shipowners back on land are wealthy in both types of the whale's value: the whale oil and spermaceti candles that light their houses (use value), and the lavish houses and parties bought with money made from selling the commodity (exchange value). Women are brought into the story of commodity circulation through their weddings: striking, in a novel almost devoid of female characters. Women function here as commodities, like cetaceans, varying by the value they fetch on the marketplace: daughters equal a whale, while nieces are "portion[ed] off with a few porpoises," the homophonic "porpoise" and "portion" cementing the equivalence. There is nothing new in the idea of marriage as a capitalist institution; Jane Austen, working fifty years earlier, mastered the ultimate proto-Marxist critique of marriage. But there is special horror in the way that Melville frequently evokes women and weddings in describing the same process of commodity circulation as butchered whales. As fungible as whales, women are accorded a nonhuman status and, as such, become subject to the direct violence of extractive capitalism.

WHALE INFRASTRUCTURES

Ishmael's account of the whaling industry radically changes as he narrates from the deck of the *Pequod*, where whale-oil-magic is disenchanted by the totalizing violence of whale slaughter. The violence is multidirectional: human whalemen slaughter whales, and some of those same whalemen are attacked by whales and sharks or are drowned. Those whalemen who survive the midocean encounter with whales are covered in the blood and gore of dying whales, events that Ishmael describes with unrelenting detail: "The red tide now poured from all sides of the monster like brooks down a hill. His tormented body rolled not in brine but in blood, which bubbled and seethed for furlongs behind in their wake."[20] Then, back aboard ship, the crew wallows in that whale's blood as they butcher and render its body for oil.

Melville might have portrayed the incredible violence of whaling as an alien phenomenon, a process utterly unrelated to the lives of readers on land. To have done so might have exonerated readers and whale oil consumers on shore from any crimes committed in their name. But while the *Pequod*'s crew dismembers each slaughtered whale, Ishmael imagines part of the whale's body in terms that would be familiar to any landbound reader: in metaphors that compare the whale's body to mundane, quotidian built environments and infrastructures back on land. Ishmael portrays the whale's anatomy by comparing it with the images of country roads and city streets, tunnels, canals, horse carts, urban waterworks, subsurface water and gas pipes, building timbers. Through the figure of the whale, in its disassembly, Ishmael explores the concept that is coming to be called energy: a concept that encompasses fuel resources along with the extensive infrastructures and ideologies that they both require and make possible.

The infrastructural metaphors in *Moby-Dick*'s cetology chapters are part of Melville's practice as an energy archaeologist. The metaphors register the intellectual and infrastructural work that accompanies violent resource extraction: the ideological abstraction of "work" and "mobility" from the violently extracted material of fuel, and the construction of an accompanying infrastructural system to sustain that vision. "Energy," as it emerges in the mid-nineteenth century, is an assemblage of natural resource, ideology, and infrastructure. Historian Christopher Jones, writing about energy in the late nineteenth-century United States, observes a similar relationship between resource extraction and infrastructure development that *Moby-Dick* theorizes here. According to Jones, fossil fuels did not simply change the operation of discrete machines and technologies; fossil fuels also reshaped the world's landscapes and waterways, as well as the possibilities for moving through them. New road, rail, canal, pipeline, and wire infrastructures were constructed to meet the needs of transporting energy, creating what Jones calls "landscapes of intensification."[21] *Moby-Dick*'s infrastructural metaphors register and perform the "social" and "cultural" work of intensification, as Jones defines it.

Metaphors are often eschewed in environmental humanities scholarship, and with good reason. In her call for a practice of oceanic studies attentive to the material realities and lived experience of maritime laborers, Hester Blum begins with an urgent admonition: "The sea is not a metaphor."[22] A resistance to metaphor runs, too, through work from infrastructure studies in anthropology and media studies, and even in scholarship from the recent infrastructural turn in literary studies.[23] But the concept of energy is one that derives its power from ideological patterns as well as material sources, and metaphors open a window into those ideological patterns. *Moby-Dick*'s infrastructural metaphors help us understand energy at its birth. They help us understand how (capitalist, settler-colonial) energy production follows an intensifying pattern that looks like this: energy resource→fantasy of land and life transformed→infrastructure development→increased need for energy resource. *Moby-Dick*'s infrastructural metaphors hasten these positive feedback loops and cement the links in the energy-infrastructure-ideology assemblage. Metaphors do this by yoking together acts of extraction at sea, where the *Pequod*'s crew slaughters whales and butchers their bodies, with infrastructures that facilitate mobility back on land. Jeffrey Insko writes that "the logic of extraction mirrors the material/epistemological logic of coloniality: *extraction* (or extractivism) is coterminous with *abstraction*."[24] Metaphors are some of the figures that carry out the logics of extraction. Metaphors extract by removing objects from their context, directing attention away from the scene at hand to a far-flung object of comparison. *Moby-Dick*'s infrastructure metaphors flee the scene of violence, but those metaphors also tether land-based infrastructures to the bloody violence of whaling. It is tempting to see the violence of energy infrastructures only in their spectacular failures: in, for example, the Deepwater Horizon oil rig explosion and catastrophic oil spill in 2010. But *Moby-Dick*'s infrastructural metaphors reveal the violence inherent in energy infrastructures and in their boring, everyday processes: as Winona LaDuke and Deborah Cowen write, "not in the system's failure but in its smooth operation."[25] Understanding how the fantasies of energy infrastructure take shape in text as metaphors could aid in an understanding of energy that ultimately breaks that chain.

The comparisons between the whale's body and everyday infrastructures begin in the Extracts, where our Sub-Sub-Librarian extracts a passage from *Paley's Theology* that compares the whale's heart to civic waterworks: "The aorta of a whale is larger in the bore than the main pipe of the water-works at London Bridge, and the water roaring in its passage through that pipe is inferior in impetus and velocity to the blood gushing from the whale's heart."[26] The comparison of the whale's heart to the London waterworks renders the whale as civic infrastructure that is both quotidian and spectacular. When the London Bridge waterworks were established in the sixteenth century, they were celebrated as technological marvels and feats, and they have been remembered by historians as a catalyst for the city's urbanization and modernization.[27] A passage in chapter 86, "The Tail," compares the three layers of the whale's tail to the layered walls of ancient Rome: "To the student of old Roman walls, the middle layer will furnish a curious parallel to the thin course of tiles always alternating with the stone in those wonderful relics of the antique, and which undoubtedly contribute so much to the great strength of the masonry."[28] Here Ishmael mimics a pedantic classical archaeologist, praising the craftsmanship of antique infrastructures.

Not all of Ishmael's infrastructural metaphors are ancient. Chapter 85, "The Fountain," compares the whale's respiratory system to modern infrastructures of intercity and intracity transport such as canals and plumbing pipes. The links between the whale's body and quotidian infrastructures like these affirm the enmeshment of everyday systems within the brutal violence of whaling. "Furthermore, as his windpipe solely opens into the tube of his spouting canal, and as that long canal—like the grand Erie Canal—is furnished with a sort of locks (that open and shut) for the downward retention of air of the upward exclusion of water, therefore the whale has no voice."[29] The whale's respiratory valve is likened to a lock that controls the depth of canals and rivers and in so doing regulates the flow of ship traffic. The whale's spouting canal is compared to the "grand Erie Canal," a strange but typically Melvillian conflation of the old world Grand Canal in Venice and the new world Erie Canal: a conflation that zips the reader's imagination across the Atlantic and through several centuries in a single breath. Melville's metaphors imaginatively render the whale's body into an encyclopedic history of terrestrial infrastructures.

The infrastructural metaphors often slip; referents transform midpassage and direct the reader's imagination across different scales of time and space. In the passage describing the whale's respiratory channel, Ishmael likens the channel first to the "grand Erie Canal." A few lines later, the respiratory channel is compared to the gas and water pipes running under city streets. "Now, the spouting canal of the Sperm Whale, chiefly intended as it is for the conveyance of air, and for several feet laid along, horizontally, just beneath the upper surface of his head, and a little to one side; this curious canal is very much like a gas-pipe laid down in a city on one side of the street. But the question returns whether this gas-pipe is also a water-pipe; in other words, whether that exhaled breath is mixed with water taken in at the mouth, and discharged through the spiracle."[30] The scale of the referent has changed; the "spouting canal" has slipped from the intercity "grand Erie Canal" to the intracity plumbing of gas and water pipes.[31] Unlike the comparison of the whale's heart to a sixteenth-century waterworks, and its tail to ancient Roman walls, Ishmael casts the whale's interior structure in the imagery of plumbing infrastructures recognizable to mid-nineteenth-century readers (as well as to twenty-first-century readers). The reference to gas pipes marks one of several instances in the book where fossil fuels are referenced, and these moments demonstrate that animal energy is coexistent with the fossil fuel regime of natural gas — as, of course, it was, in spite of a conventional narrative of energy history that sees fossil fuel technology as a radical break with the past.

The metaphors that liken the whale to contemporary infrastructures portray the whale, not as alien or preliminary to infrastructures called modern, but as consonant with them even at the level of anatomy. A whaling booster might have limited the metaphorical comparisons of the whale to contemporary infrastructures only, trying to establish the whale's nativity to the contemporary nineteenth-century economy that its body served. But Melville presents a critical archaeology of infrastructures, situating the whale within ascendant technologies and political powers as well as within structures of the past, the obsolescent, the forgotten, and the arcane. The strategy of linking the anatomy of the whale to quotidian terrestrial infrastructures is part of how the novel establishes the connection between the primeval and uncanny whale's body and the emerging fossil-modern world on land, between the soon-to-be-obsolescent regimes of whale oil and the regime of fossil fuels to come, and between the thriving industries and technologies of the present and the ruins of the past.

A taxonomy of the metaphors used to describe whales in *Moby-Dick* could be made on a huge number of subjects — not only infrastructure.[32] But the infrastructural metaphors in *Moby-Dick* are worth reflecting on because infrastructure, mobility, mechanization, and urbanization are the material purposes that the whaling industry itself served. The whale aided the American projects of territorial expansion, economic growth, and infrastructure development materially as well as metaphorically. Whale oil lubricated factory machinery, lit city streets, and helped create the market that petroleum would explode into.[33] Ishmael does not limit himself to imagining the specific, material purposes to which whale products will be used to advance infrastructure development back on land. The imaginative leap that Ishmael makes between the whale's body and civic waterworks, city walls, plumbing and gas pipelines is characteristic of the rendering imagination: an instrumentalist way of seeing whales as resources bound to terrestrial human structures rather than as beings unto themselves. Ishmael's infrastructural metaphors show just how many systems and structures in the world are connected to the animal capital of the whale. At the moment of the birth of the concept of energy, *Moby-Dick*'s infrastructural metaphors model the way that an energy resource, the whale's body, is put to use in terrestrial infrastructures: first in the imagination and then, fatally, in practice.

WHALE INFRASTRUCTURES AND SETTLER COLONIALISM

Through his disquisition on whale parts, Ishmael also accounts for the specific infrastructures and processes of settler colonialism and, in particular, the settlement of the North American West in the nineteenth century: the western movement of white settlers, the genocide of Indigenous people, and the engineering of land for cities and commodity agriculture. Colonialism, resource extraction, and infrastructure development are mutually co-constitutive practices, as countless postcolonial and Indigenous scholars and activists have long argued.[34] The whale's body is a multivalent figure for settler colonialism: sometimes the perpetrator of imperial violence and sometimes the object of it.

The intensification of setter colonialism through energy infrastructure is evoked throughout *Moby-Dick*, fittingly enough, through images of and metaphors for the whale's mouth and jaw, which stand for consuming hunger. The invocations of settler colonialism located inside the whale's mouth evoke Audra Simpson's characterization of Mohawk people "operating in the teeth of Empire, in the face of state aggression."[35] The infrastructural images attached to the whale's mouth also perfectly illustrate the type of infrastructures that Winona LaDuke and Deborah Cowen call "Wiindigo infrastructure," after the all-consuming, cannibal monster in Anishinaabe belief.[36] The whale's connection to colonial expansion is not only metaphorical, either. In chapter 74, "The Sperm Whale's Head—Contrasted View," Ishmael explains how sailors on board the ship will craft the jawbone into consumer goods for friends back home: "canes, umbrella-stocks, and handles to riding-whips."[37] The whale's jawbone aids mobility through walking (canes and umbrella handles) and riding in horse-drawn carts (riding whips).

In addition to furnishing these *material* aids to mobility and transportation, the whale's jaw also evokes through *metaphor* the settlement of new territories and movement within them. Chapter 74 ends with Ishmael's careful examination of the sperm whale's jaw and mouth. Ishmael likens the disassembly of the sperm whale's jaw and the removal of its teeth to acts of clearing land in the unsettled American West.

> With a long, weary hoist the jaw is dragged on board, as if it were an anchor; and when the proper time comes—some few days after the other work—Queequeg, Daggoo, and Tashtego, being all accomplished dentists, are set to drawing teeth. With a keen cutting-spade, Queequeg lances the gums; then the jaw is lashed down to ringbolts, and a tackle being rigged from aloft, they drag out these teeth, as Michigan oxen drag stumps of old oaks out of wild wood-lands. There are generally forty-two teeth in all; in old whales, much worn down, but undecayed; nor filled after our artificial fashion. The jaw is afterward sawn into slabs, and piled away like joists for building houses.[38]

At first, the whale's jaw metaphorically becomes land that needs to be cleared of giant old tree stumps: remnants of an Indigenous ecosystem are figured here as obstructions to settlement. The harpooneers are characterized both as "accomplished dentists" and "oxen" in this passage, committing simultaneous acts of extraction. The metaphor slips as the passage continues: the jaw, which in the tree-stump analogy had been the land, transforms as it is sawed apart and disassembled into the building materials for a house. The morphing metaphor of the whale's jaw, as it slips from land to building material, becomes a narrative of land settlement and settler colonialism: violent resource extraction and ecosystem destruction followed by the erection of houses for new occupants. It is striking that the Indigenous harpooneers are the laborers who perform the labor of extraction; their role in metaphorical settler colonialism is another instance of the upturned racial hierarchy aboard the *Pequod*. It is not only Melville who saw settler colonialism in a whale; the body of the whale in the age of US whaling was a frequent figure for settled land. A children's chapbook published by Rufus Merrill and contemporaneous to *Moby-Dick* presents the whale as a synecdoche for settled land: "His [the whale's] mouth is 30 feet long, and, when open, it is large enough to drive in a horse and wagon."[39]

Moby-Dick's description of settler colonial resource extraction echoes some Indigenous critiques of resource extraction, which have centered in recent years on critiques of fossil fuel extraction and, in particular, on resistance to energy infrastructure projects such as the Keystone XL and Dakota Access Pipelines.[40] Nick Estes, a citizen of the Lower Brule Sioux tribe, Potawatomi scholar Kyle Powys Whyte, Sisseton Wahpeton Oyate scholar Kim TallBear, Métis scholar Zoe Todd, and others explain that US energy independence—the pretext for settler energy infrastructure such as pipelines—is in direct opposition to the sovereignty of Indigenous nations and the well-being of Indigenous people and lands.[41] As Kyle Powys Whyte puts it, "Settler colonial injustice is environmental injustice," and the reverse is often true as well.[42] The US commercial whalers making commodities out of whale bodies and the settlers with their oxen pulling tree stumps from forested land in Michigan are engaged in the same project of environmental violence and Indigenous genocide and displacement as white US oil developers who perpetually commit violence against Indigenous people: for example, twentieth-century oil prospectors in Kansas and Oklahoma who exploited and even murdered Indigenous people in order to access oil reserves on Cherokee and Osage lands, and twenty-first-century oil companies plotting the Dakota Access Pipeline through Standing Rock Sioux lands or the Enbridge Line 3 through Anishinaabe lands.[43] The US commercial whaling industry was an instrument of many forms of colonial violence: the whaling industry appropriated and exploited Indigenous labor, provided material aid to settlers in the Pacific Islands, and, in the case of US commercial whalers in the Bering Strait, depleted the bowhead whale populations, which removed a vital source of sustenance and energy for Indigenous people who live there.[44]

The infrastructural metaphors in *Moby-Dick* lay bare the relationship between resource extraction, infrastructure, and Indigenous genocide and displacement. But there are limits to thinking about Melville as a writer in solidarity with Indigenous writers and activists. Melville renders the violence of whale slaughter in unflinching detail, but he does indeed flinch when it comes to representing violence against Indigenous people, whose existence is only implied in scenes from the settlement of Michigan or in comparisons of the whale's jaw to an Indigenous dwelling such as a tipi or wigwam. Through the settler ideology of the "vanishing Indian," Indigenous genocide is euphemized as a natural "extinction" and relegated to the ancient past in the passage explaining the *Pequod*'s name: "*Pequod*, you will no doubt remember, was the name of a celebrated tribe of Massachusetts Indians, now extinct as the ancient Medes."[45]

Melville's treatment of Indigenous people is doubly blind: blind to the presence and specific languages, cultures, or identity of Indigenous people on board the *Pequod* (and, by extension, the many Indigenous laborers in the nineteenth-century US commercial whaling industry) and blind to the actual violence of settler colonialism throughout the United States and its expanding territories. The commercial whaling industry was populated by a large number of Indigenous whalemen, and commercial whaling was also a huge presence, and source of subsistence, for Indigenous communities from the East Coast. Historian Nancy Shoemaker argues that, in fact, most young men from Indigenous communities in southern New England worked on commercial whaling ships in the eighteenth and nineteenth centuries.[46] Tashtego, the Gay Head harpooner hailing from Martha's Vineyard, stands in for many of these Indigenous laborers. As a skilled whaleman who has stalled out in the position of harpooner rather than achieving higher leadership positions on board the ship, Tashtego resembles many Indigenous whalemen who were kept out of positions of leadership.[47] But he is also a cardboard character. The Indigenous crew members on the *Pequod*—Queequeg, Tashtego—are notable for their harpooning skills, but their attitude toward whales as commodities to be extracted is not differentiated from that of the settler whalers on board. Melville reserves descriptions of violence for the whales themselves, only squinting at the violence of settler colonialism.[48] The novel similarly evades the violence of slavery, evoking enslaved people in the American South through the proxy of whales in chapter 89, "Fast Fish and Loose Fish." The violence of settler colonialism and white supremacy are sublimated through infrastructural imagery and metaphors that are paradoxical in their effect. On the one hand, infrastructural imagery marks the whale's body as extraordinarily large, less a discrete body than a network, with the potential for movement and mobility within it. On the other hand, the infrastructural imagery used to describe the whale makes it invisible: the unremarkable stuff of everyday life in walls, water, and pipelines. That infrastructure, resource extraction, and—indeed—violence itself are often invisible does not mean that they are unimportant, only that they are so pervasive as to be taken for granted.[49]

Put simply, Melville cannot imagine nonextractive ways of being with whales or nonhuman beings. Commercial whaling as portrayed in *Moby-Dick* is death-driven settler ideology that makes whale bodies into commodities and infrastructures that fuel capitalist expansion and settler colonialism. There are countless other ways of imagining whales and even whaling. Iñupiaq whalers provide another way of imagining being-with-whales. In traditional Iñupiaq knowledge, whaling is an exchange in which discerning, attentive whales give themselves to deserving hunters.[50] In a brilliant essay that rewrites *Moby-Dick*, "Citation in the Wake of Melville," contemporary Iñupiaq poet Joan Naviyuk Kane describes hunting as a life-giving practice completely outside the system of extractive capitalism: whale, seal, and walrus bodies gave her family *life*, not capital or commodities. Kane reiterates Melville's caricature of Indigenous people in order to satirize his work. Kane quotes Melville on the humans who eat whale flesh: "Only the most unprejudiced of men like Stubb, nowadays partake of cooked whales; but the Esquimaux are not so fastidious. We all know how they live upon whales, and have rare old vintages of prime old train oil. Zogranda, one of their most famous doctors, recommends strips of blubber for infants, as being exceedingly juicy and nourishing."[51] Kane follows up the Melville quotation with her own family's history of whaling and other subsistence hunting practices. Like Melville, Kane does not shy away from the violent facts of animal hunting: "We used to live on whales. On polar bear soup (should you, lacking, say, a *tuukaq* or a gun, ever need to kill a polar bear, simply create a pronged baleen cage around a scrap of anything scented like prime old train oil, leave it somewhere a polar bear can find it and devour it, and the cage ought to open in the polar bear's stomach, causing eventual rupture and death). On the aforementioned walrus. On seal."[52] But Kane's description of Iñupiaq hunting does not end with violence or by describing how animal bodies are rendered materially and metaphorically into settler energy and commodities. While Melville writes about how commercial whalers *die* on whales, Kane writes about how "we used to live on whales," portraying whaling as a sustaining and reciprocal cycle of life. She even interrupts her own family history of hunting to describe how animal flesh literally sustains her family in the present: "—A digression here to say that last week I prevented the death of one of my mother's siblings, my uncle. He's still in the hospital. But before my husband and I carried him to the emergency room, I indulged his appetites in my kitchen and fed him half-dried seal soaked in seal oil. It was exceedingly juicy and nourishing."[53] Kane repeats Melville's language "exceedingly juicy and nourishing." But where Melville condescendingly attributed the lines to a specious "Esquimau" doctor in order to mock Indigenous knowledge, Kane gives them sincerely and in good faith, turning that mocking condescension back on Melville and settlers like him who do not know how to sustain life together with whales and other marine creatures.

Moby-Dick AND THE DENIALISM OF PETROLEUM CULTURE

So far, I have explored the imagination of growth in extractive industry: the enticing illusion of whale-oil-magic in New Bedford, and the promise of infrastructural and imperial growth imagined through metaphors in the whale's body. But Melville's imagination of growth is double-edged:

growth in extractive industry is endless and also intolerable.⁵⁴ I now explore *Moby-Dick*'s capacious imagination of the limits to growth and, eventually, whaling's end. Most scholars of *Moby-Dick* have focused on the novel's portrayal of extinction as the ultimate ending. John Levi Barnard, for example, argues that whaling and other industries based on the commodification of animal flesh are "constitutive elements of an *extinction-producing economy*."⁵⁵ *Moby-Dick*, argues Barnard, is aligned with his own critique of animal capital, "casting a critical eye on a global consumer economy and a culture of consumption of nonhuman animals that together inexorably undermine their own ecological viability."⁵⁶ Other extinction-oriented critics focus on the explicit invocation of extinction in chapter 105, "Does the Whale's Magnitude Diminish?—Will He Perish?" The chapter develops by reflecting on how whale behavior seems to be changing in response to the threat of hunting—a tacit acknowledgment of serial depletion if not exactly of extinction—and concludes with an affirmation of faith in the whale's ultimate protection from extinction: "We account the whale immortal in his species, however perishable in his individuality."⁵⁷ Some have read the passage as Ishmael's dismissal of extinction anxiety; others, including Barnard, credit Melville with confronting extinction and a posthuman future.⁵⁸ *Moby-Dick* was a rare nineteenth-century whaling text in its exploration of extinction; widespread anxiety about whale extinction would arrive in whaling culture much later. Some observers in the late nineteenth and early twentieth centuries cited species extinction as a risk of so-called modern whaling with mechanized harpoons and bomb lances—although as I explore in chapter 4, species extinction often served these observers as a cover for nativist concerns about the changing labor force in the whaling industry. Whale extinction did not become a major political concern until, arguably, the Save the Whales movement of the 1960s.⁵⁹ But extinction is only one way of understanding the limits of an extractive industry like whaling.

Moby-Dick also confronts the end of whaling through the forms that energy culture gives to limits: finitude, depletion, resource peaks, and exhaustion. Anxiety about limits is coeval with fossil fuel extraction.⁶⁰ The culture of underground mineral extraction has an older and arguably wider range of ideological resources for describing the anxiety of exhaustion than whaling culture does, perhaps because mineral resources are so patently finite. *When will the oil run out, and what will we do then?* Writing about extraction in Victorian literature, scholar Elizabeth Miller acknowledges the major-key "voice of optimism and progress" in extraction literature, but her book "tunes into [the] minor key" of extraction culture: "the ever-present sense that [industrial Britain] was living on borrowed time."⁶¹

In so doing she finds that "the mood of finitude, of removing something that is irreplaceable and subject to looming environmental limits, pervades extraction ecology."[62] The "mood of finitude" that Miller associates with mineral extraction also pervades *Moby-Dick*, and many of the forms used to describe the finitude of mineral resources applies also to the forms used in *Moby-Dick* to describe the ending of the industry. *Moby-Dick* is also an effective key to other texts of energy culture, precisely because it plays in the major key of optimism and, at the same time, the minor key of finitude. The novel parrots the delusions and denials that drive energy extraction, and it forces its reader's attention on the many limits and possible endings of extractive capitalism. *Moby-Dick* is resolutely *about* whaling, but as this chapter shows, its forms are often drawn from, and applicable to, other types of energy extraction.

To read *Moby-Dick* through the lens of mineral and fossil fuel culture is also historically fitting, given the novel's emergence within the same century as many revolutionary geological discoveries: notably the new theory of geological time, driven by fossil discoveries, that radically expanded scientists' understanding of the planet's age.[63] Other critics have noted evidence of Melville's geological imagination throughout his writing, and especially in *Moby-Dick*, where Ishmael's imagination ranges across the newly understood expanse of deep geological time.[64] The movement in *Moby-Dick* from chapter 104, "The Fossil Whale," to chapter 105, "Does the Whale's Magnitude Diminish?—Will He Perish?" represents an oscillation in geological time from the deep fossiliferous past to the distant future that is characteristic of nineteenth century geological accounts.[65] Most historians of geology writing about the geological revolutions of the nineteenth century write about the new geologic timeline and discoveries in the fossil record, drawing on accounts by eminent geologists such as Charles Lyell and Georges Cuvier.[66] But geology is not only a disinterested science. New techniques for the mass extraction and refinement of coal and petroleum were also significant discoveries in geological science and the subject of a huge body of professional and popular literature in the nineteenth century. These types of geological writings receive less attention in historical accounts because they are so thoroughly entangled with the business of oil extraction. But these texts, too—vernacular accounts and speculations written by oil laborers, by geologists and chemists who sought to make sense of petroleum and its applications, and by journalists who covered the oil booms of the nineteenth century—are an important context for *Moby-Dick*. Placing *Moby-Dick* in the company of nineteenth-century geological research and petroleum narratives puts new light on the novel's orientation toward scarcity and environmental limits. Geological texts written by and for the coal and oil industries give voice to the business booster's optimism about future prospects, but they, too, are haunted by the mood of finitude. They promise endless future prospects for industry, but they deliver those promises in haste, as if stifling a nervous laugh. Melville's exuberant novel stifles nothing, neither the feverish excitement of prospectors nor the shock of violence nor gloom of extinction.

Nineteenth-century oil narratives deployed contemporary geological science in order to ask, "Whence came oil?" and "Where is it?" But their most pressing anxieties were: "Whither oil?" and "Will oil eventually be exhausted?" Oil field chronicler Edmund Morris considered the question in a chapter of his narrative, *Derrick and Drill; or, An Insight into the Discovery, Development, and Present Conditions and Future Prospects of Petroleum* (1865). Morris begins his query with the observation that the yield of oil wells in Pennsylvania has decreased by half in the three years between 1862 and 1865—a phenomenon Morris describes as "at the first glance, quite inexplicable upon any other ground than that the supply of oil is becoming exhausted."[67] Morris continues by reviewing the opinions of various experts concerning the likelihood that oil will be exhausted. But he answers his anxiety with a series of tormented logical leaps. First, Morris quiets his "terror" about exhaustion by contradicting recent evidence and his own direct observation by speculating that future advances in technology will solve the problem and make exhausted wells produce again. Then Morris deploys stories and archaeological evidence of oil extraction from such places as Babylon, Greece, Egypt, Baku, and North America before settlement to claim that if oil has not yet been exhausted, then it will not be exhausted in the future: "The oil regions of the old world having continued to furnish constant supplies for many centuries, the presumption with many is that those of this country will be found equally inexhaustible."[68] Morris completely ignores the question of scale; he anticipates the potential benefits of a coming Great Acceleration in resource consumption but does not consider the consequence that the resource might be exhausted.

Another nineteenth-century popularizer of geological science, William Denton, dismissed fears of oil exhaustion through an illogical analogy of oil to water, which Denton views as inexhaustible:

"But what shall we do," says a forward-looker, "when the water of the earth is all burned, the earth cold, the coal gone, and the oil consumed?" We shall never meet with such disaster; for this reason,— when hydrogen and oxygen gases, of which water is composed, are burned, water is produced by their combustion in exact proportion to the amount of the gases used: so that it might be burned over and over again forever. As long as the world exists, then, we may be assured that man's ingenuity will keep pace with his necessities of this kind, and the human race march on to the goal that shall lie before them.[69]

In this passage, Denton views the planet's geological history as a simplistic chemical reaction in which nothing can be lost and no real, lasting harm is possible: these laws exist in direct contradiction of emerging thermodynamic science, which grapples with the problem of waste in energy transformations. These passages about the futurity of oil from Morris's and Denton's narratives represent a wider pattern in oil narratives and popular geological histories of the nineteenth century: they voice and consider the eventual depletion of oil but ultimately dismiss that concern.

As this sampling of nineteenth-century oil texts indicates, the problem of resource finitude was met with sweaty, illogical claims. *Moby-Dick* resembles these oil narratives as Ishmael follows the same illogical arc in his dismissal of the threat of whale extinction. Ishmael states the likelihood of resource depletion but eventually and confidently follows it up by vanquishing the concern in chapter 105's final paragraph: "Wherefore, for all these things, we account the whale immortal in his species, however perishable in his individuality. He swam the seas before the continents broke water; he once swam over the site of the Tuileries, and Windsor Castle, and the Kremlin. In Noah's flood he despised Noah's Ark; and if ever the world is to be again flooded, like the Netherlands, to kill off its rats, then the eternal whale will still survive, and rearing upon the topmost crest of the equatorial flood, spout his frothed defiance to the sky."[70] Pursuing the same narrative strategy as Edmund Morris in the oil account that uses the history of oil extraction to explain its futurity, Ishmael voices the deceptions that fed extractive capitalism in the nineteenth century: first, in spite of instances of local destruction, enduring environmental harm is impossible, and second, history serves to console us for any harm we cause in the present. These deceptions still feed extractive capitalism today.[71]

PEAK WHALE OIL

Starting in the mid-twentieth century, a new name emerged for the acknowledgment, anticipation, and anxiety of resource depletion: "peak oil." Peak oil is the name first given by geologist M. King Hubbert in the mid-twentieth century to the theoretical moment when petroleum reaches its peak production and begins to decline—not solely because of the depletion of the resource but on account of a huge number of complex factors, including technological capacity.[72] Global peak oil is a theory, not an observed (or potentially even observable) moment, but most geological and political theorists place peak oil in the future. Hubbert's theory of peak oil is in fact a euphemism for the end of oil, for what inevitably follows a peak is decline and, ultimately, the end. Peak oil also refers to a loosely constellated political movement dedicated to forecasting and preparing for peak oil and post–peak oil life, which will incite a widespread collapse of infrastructure, political structures, and social norms.[73] As a political movement, peak oil is apocalyptic, one in long chain of (mostly) American millennial traditions, and part of its practice is to locate and produce visualizations of the post-peak future. Mainstream traces of or touchstones for post-peak culture include such films as *The Postman*, *Mad Max*, and *Water World* and James Howard Kunstler's book *The World Made by Hand*, all of which feature postapocalyptic worlds peopled by brutal, violent leaders and scrappy communities who create their world from the salvage of the old.[74]

Moby-Dick is a work of *peak whale oil*, in that peak oil is an analogue for another type of exhaustion-anxiety that *Moby-Dick* evinces. Like peak oil, peak whale oil is only observable in hindsight; economic histories of the US industry show, in fact, that whale oil production did in fact peak in the 1850s, the decade when *Moby-Dick* was published.[75] *Moby-Dick*, too, is somewhat future-oriented in its representation of an obsolete whaling world; Ishmael ships out of Nantucket, the whaling port that had by the time of the book's publication already been surpassed by New Bedford. The *Pequod* is a salvage ship, a "cannibal of a craft" that is held together by pieces of the quarry of the hunt: whale teeth and jawbones that speak to its history, age, and the clever salvaging of the people who build the ship. And, as chapter 105 testifies, the novel is haunted by the threat of resource depletion. To read *Moby-Dick* as a peak whale oil novel is not to claim that the novel was clairvoyant; Melville could not and did not prophesize the end of the whale oil industry. Instead, he saw that the industry, like all industries of extractive capitalism, had built-in limits that were apparent to observers who did not blindly subscribe to the mandate of endless growth.

To translate the terms by which we consider the extermination of whale life from extinction to resource depletion might seem cool and bloodless, as if the whale possessed no sovereignty of life or ecological purpose beyond supplying humans with oil and other commodities made from its body. That is not my aim here. To approach whale species depletion and extinction in terms of resource exhaustion is to consider whale oil in the same terms that many consider petroleum and other fossil fuels: as resources threatened by scarcity rather than as forms of life threatened by violent destruction. The inverse is true: extraction should always be considered an act of direct violence whether or not the extracted resource is itself a creature. All regimes of extractive capitalism, whether extraction is carried out on living creatures such as whales or on nonliving substances such as oil, pose an existential threat to human and nonhuman life alike.

Theorizing resource peaks from the perspective of whale oil rather than petroleum reveals the inherent violence in extraction, which is never not bloody. The case of whaling also shows the ways in which theories of resource peaks fail and fall short. The slaughter of whales did not, in fact, decline when the market for whale oil declined. *Moby-Dick* might be a work of *peak whale oil*, but it is not a work of *peak whaling*; whaling's peak would come much later in the twentieth century. Petroleum and petroleum products may have obviated whale oil, but the slaughter of whales accelerated in the late nineteenth and twentieth centuries, after the peak of whale oil production.[76] It is likely that peak oil predictions are failing in a similar way. Like so many other millennialist movements that fizzled when the millennium did not arrive on schedule, peak oil as a political or social movement has quieted. On account of new extraction technologies like ultradeep drilling and fracking, recent oil production has in fact risen from a trough that many assumed was the start of oil's final decline; as Stephanie LeMenager puts it, "reports of oil's death have been exaggerated."[77] What has come in the wake of easy oil, she writes, is Tough Oil. And what happens in the era of Tough Oil is even more violent than what came before.

What these peak oil prophecy errors can do is shift our moral attention away from the problem of resource scarcity and back to life itself: the forms of violence committed in extractive industries can broaden and accelerate in the face of scarcity anxiety. That forecasts of peak oil—and peak whale oil—were made in error, without having anticipated great rates of extraction to come, does not mean that oil or any other extractive resource is infinite. What this means, in part, is that the scarcity of the resource ought to be only one part of extraction anxiety: we who live in extractive capitalism also ought to be vigilant against forms of violence that result as technological advances make extraction even more efficient and productive. The failure of peak oil forecasts is that they are limited and ethically anesthetic; resource depletion is only one part of the problem of extraction.

Moby-Dick is a peak whale oil novel because it apprehends and anticipates the depletion of the whale oil resource and because, viewed from certain angles, the novel has a postapocalyptic, salvage aesthetic that resembles latter-day peak oil fictions. But the novel has other and even more instructive ways of visualizing resource exhaustion than by imagining its peak. *Moby-Dick* characterizes exhaustion as a threat faced mutually by humans and whales. The novel evokes the mutual threat of depletion through an unlikely figure—a tobacco pipe—to which its characters turn in moments of emotional exhaustion; the pipe links extinction with exhaustion in every sense of the word.

For example, the grave, cosmic phenomenon of species extinction is given in chapter 105 through the whimsical image of a whale smoking "his last pipe": "The moot point is, whether Leviathan can long endure so wide a chase, and so remorseless a havoc; whether he must not at last be exterminated from the waters, and the last whale, like the last man, smoke his last pipe, and then himself evaporate in the final puff."[78] The image, however ludic, cements the connection between human and nonhuman extinction. Pipe imagery recurs frequently throughout *Moby-Dick*, and paradoxically, pipes are evoked at moments of a whale's violent death as well as when human characters are trying to calm themselves. Second mate Stubb is characterized by the pipe he constantly has in his mouth, and Ishmael accords Stubb's pipe with the power of keeping Stubb's disposition sanguine and carefree; it was a tool that protected him "against all mortal tribulations."[79] Stubb keeps calm when he confronts the whale, and it is he who kills the first whale of the *Pequod*'s voyage. Ahab smokes a pipe, too, but early in the narrative, he throws his pipe away. Addressing the small object before tossing it overboard, Ahab apostrophizes: "Oh, my pipe! hard must it go with me if thy charm be gone! Here have I been unconsciously toiling, not pleasuring,—aye, and ignorantly smoking to windward all the while; to windward, and with such nervous whiffs, as if, like the dying whale, my final jets were the strongest and fullest of trouble. What business have I with this pipe? This thing that is meant for sereneness, to send up mild white vapors among mild white hairs, not among torn iron-grey locks like mine. I'll smoke no more—."[80] Ahab is rattled because in puffing nervously on his pipe, he suddenly reminds himself of a dying whale making its last, labored exhalations. And, as mentioned, the whale in chapter 105 "smokes his last pipe" at his death and—because he is an endling, the last living individual of a nearly extinct species—at the moment of his entire species' extinction.

Viewed within this context, Ahab's last pipe, which he smokes near the beginning of the *Pequod*'s cruise, is stark foreshadowing not only of his own death but of a massive extinction that encompasses both humans and whales. Pipe-smoking signals the affect of exhaustion: an affect not of overwhelming grief but of calm, almost narcotic, resignation. Through the pipe and other images, *Moby-Dick* characterizes whaling through the paradigmatic voyage of the *Pequod* as a teleological progression to the mutual destruction of whales and humanity. I am not the first to note the significance of pipe imagery; critics Elizabeth Schultz and John Levi Barnard have both discussed how the pipe in *Moby-Dick* links humans and whales in shared extinction.[81]

But the other critics who have noted the pipe-smoking extinction imagery in *Moby-Dick*—and indeed Melville himself—all stop just short of explaining the implications of pipe-smoking whales and humans confronting their extinction together. Smoking pipes together makes whales and humans kin. Ishmael and Queequeg share a pipe at the Spouter Inn when they cement their relationship, when Ishmael feels a "melting in me" and Queequeg declares them "married": "If there yet lurked any ice of indifference toward me in the Pagan's breast, this pleasant, genial smoke we had soon thawed it out, and left us cronies."[82] And each in his own bloody, misguided way, Ishmael and Ahab yearn for intimacy with their whale kin: wounded Ahab by pathologically punching through the "pasteboard mask" of the whale's alterity, and schoolmasterly Ishmael through his obsessive cetology that links whales to the human world through endless histories and metaphors. It is through the pipe imagery that the novel comes closest to acknowledging the kinship between whales and humans. But the novel and its characters, embedded as they are in settler, capitalist epistemologies, never find language of kinship except through the humble little pipe that passes between whales and humans. Through the pipe imagery, there is a glimmer of kinship, and a world beyond extractive relations. Resource extraction is a violation of the responsibilities and privileges of kinship: care, mutual support, love. Métis scholar Zoe Todd writes about the way that Indigenous epistemologies are predicated on the kinship of the human and the more-than-human world; Todd's research focuses on what she calls "fish kin." In an essay prompted by an oil spill in the North Saskatchewan River in 2016, Todd confronts her orientation to the more-than-human substances that were reported to have killed her fish kin: the petroleum and chemical diluents that spilled into the river. These substances, too, are kin, Todd writes: "fossil kin," when approached through a kin's curiosity and care, appear not as evil substances in and of themselves but as carbon beings that have been "weaponized through petro-capitalist extraction."[83] Thinking of oil as fossil kin challenges Todd to "mobilise those aspects of Métis law that I grew up with in the service of imagining how we may de-weaponise . . . oil and gas."[84] Kinship with more-than-human beings changes the relationships that bind extractors and extracted bodies together and opens up new alternatives to extraction, new futures altogether. Not for the crew of the *Pequod*, however; Ahab has thrown his pipe away.

BUILT-IN OBSOLESCENCE

Moby-Dick falls short of imagining alternative relationships between human and more-than-human beings. But within the imaginary confines of extractive capitalism, the novel offers a keen reflection on the future of energy economies after energy transition: obsolescence, the sister anxiety of extinction and resource scarcity. Obsolete technologies, infrastructures, tools, artifacts, labor practices, and even people pile up at moments of energy transition. They do not vanish in the magic trick of creative destruction.

Ishmael makes himself a student of energy transition by observing the transitions *within* in the history of US commercial whaling. From his perspective at the peak of the US whale oil industry, Ishmael subdivides the history of whaling into the bygone Nantucket era and the present-day New Bedford era. At the time of *Moby-Dick*'s creation in the 1840s and its publication in 1851, New Bedford was indeed the center of a thriving industry. The island port of Nantucket, by contrast, was well into its obsolescence as a whaling port. Most whaling voyages had shipped from Nantucket in the early nineteenth century, but the whaling industry quickly centralized and expanded in New Bedford, likely because of the natural advantages of its protected harbor and its easy connections to coastal and overland oil shipping for the expanding US market. Ishmael tells the history of the Nantucket and New Bedford eras when announcing his intention to sail out of Nantucket: "My mind was made up to sail in no other than a Nantucket craft, because there was a fine, boisterous something about everything connected with that famous old island, which amazingly pleased me. Besides though New Bedford has of late been gradually monopolizing the business of whaling, and though in this matter poor old Nantucket is now much behind her, yet Nantucket was her great original—the Tyre of this Carthage;—the place where the first dead American whale was stranded."[85] Most first-time whalemen in the mid-nineteenth century would not pass through New Bedford in order to ship out of Nantucket. Melville, for example, did not; he shipped out of New Bedford. But Ishmael's reasons for shipping out of Nantucket are strange for several reasons beyond their plausibility. Ishmael's analogy of Nantucket and New Bedford to Tyre and Carthage hints at an awareness of future decline. Carthage eclipsed Tyre, but Carthage itself would be eclipsed by Rome—and Rome itself in the eighteenth and nineteenth centuries was synonymous with ruin rather than contemporary rule. So, Ishmael's budding history of the whale oil industry, from within that industry's peak prosperity, was already fraught with the anticipation of coming doom.

To Ishmael, Nantucket, a crumbling port dispossessed of its whaling wealth, was a window onto the past. The architecture of Nantucket, and even its inhabitants were old, their habits strangely arrested in time. The Quakers of Nantucket are described not only as alien or other to Ishmael and Queequeg but as explicitly old-fashioned. Melville writes that Nantucketers "*retain* in an uncommon measure the peculiarities of the Quaker."[86] The word "retain" signals that the scriptural language of the Quakers is not merely strange but *vestigial*, a relic of the past persisting unnaturally into the present. "So that there are instances among them of men, who, named with Scripture names—a singularly common fashion on the island—and in childhood naturally imbibing the stately dramatic thee and thou of the Quaker idiom, still, from the audacious, daring, and boundless adventure of their subsequent lives, strangely blend with these unoutgrown peculiarities, a thousand bold dashes of character, not unworthy a Scandinavian sea-king, or a poetical Pagan Roman."[87] The character of the Nantucket Quaker, says Ishmael here, is a combination of bold adventurousness—making him equal to the sea kings of Scandinavian sagas or the dramas of ancient Rome—and a mannered, archaic verbal idiom that makes his language seem better suited for theater than real life. Both of these essential elements (the adventurousness and the idiom) are described as belonging to the deep past of scripture, saga, and classical drama.[88] In locating the Nantucketers' "unoutgrown peculiarities," *Moby-Dick* writes the outline of a particular myth of the American past. Melville's neologism "unoutgrown" encapsulates the conventional *expectation* (that most people outgrow the scriptural language of the past) and its *preclusion* (that the Nantucket Quakers still speak that supposedly archaic tongue).

Ishmael's hipster-like obsession with the authenticity of old things led him to choose to ship on an obsolescing ship within the obsolescing port of Nantucket.[89] About the *Pequod* Ishmael exclaimed: "You never saw such a rare old craft as this same rare old *Pequod*. She was a ship of the old school, rather small if anything; with an old fashioned claw-footed look about her. . . . A noble craft, but somehow a most melancholy! All noble things are touched with that."[90] The *Pequod*'s leading characteristic is its age; the word "old" is repeated four times in this short passage. The *Pequod*'s age produces a prismatic array of nostalgic affects: veneration for the "noble" craft; condescension for the "claw-footed" vessel, materially diminished through metaphor to a bathtub or a piece of household furniture; melancholy for the passage of time embodied in this floating ruin. The ship is introduced with another historical comparison along the lines of the earlier Nantucket/Tyre analogy: "*Pequod*, you will no doubt remember, was the name of a celebrated tribe of Massachusetts Indians, now extinct as the ancient Medes."[91] As most readers will readily recognize, the myth of the vanishing Indian is at work here; the Mashantucket Pequot tribe is not extinct but in fact a federally recognized tribal nation in Connecticut.[92] In evoking Indigenous "extinction," Melville does not stand outside the settler practice of what Jean O'Brien calls "lasting," a historical technique for "writing Indians out of existence."[93] Like the Nantucket/Tyre analogy, the link between the ship *Pequod* and the supposedly "extinct" Pequot people is a practice of history and, at the same time, an anticipation of coming doom through imperial conquest.

Like Nantucket at large, the *Pequod* is old and old-fashioned, if not yet obsolete. The ship itself embodied a specimen of shipbuilding that would stand out on the docks of any whaling port as old-fashioned and, beyond that, simply old. Critics eager to find traces of Melville himself in Ishmael sometimes liken the *Pequod* to the *Acushnet*, the ship on which Herman Melville sailed in January of 1841.[94] But the *Acushnet* and the *Pequod* were vastly different ships. Melville shipped on the *Acushnet*'s maiden voyage, and the ship was large, new, and up-to-date in its design and technology. Although all sailing ships wear an antiquated appearance to observers today, the *Acushnet* to an observer in 1841 would have appeared a specimen of crisp new technology and impressive size. The *Pequod*, by contrast, was marked as antiquated not only by its trophies and worn surfaces but by its size and structure. A ship like the *Pequod* even in mint condition would, by the 1840s (when Melville went whaling) and even more so by the 1850s (when *Moby-Dick* was published), be regarded as an outdated ship. Even in Melville's time, whaling ships lasted longer than other types of ships and so often presented an antiquated appearance to observers, especially when viewed alongside naval or clipper ships. Whaling ships lasted longer because they were rigorously repaired and rebuilt after years-long voyages, whereas merchant ships stayed in port for as little time as possible.[95] Even in an environment where many whaling ships appeared old-fashioned, the *Pequod* appeared particularly old and obsolete. First, the ship is small, a "ship of the old school, rather small if anything": the size itself signaled its great age and limited productivity.[96] Throughout the period in which American shipbuilders were designing and building ships specifically for whaling (roughly throughout the nineteenth century), vessel size increased. Larger ships accommodated longer voyages and greater oil storage capacities, and agents and shipowners found that longer voyages were more productive and profitable and that they minimized risk: a larger single ship could bring back more oil while hazarding the risks of only a single voyage.[97] A single whaling voyage was an expensive and risky venture, and maximizing the profits of each single voyage was crucial. The size-to-age ratio (bigger equaling newer) would have been an unconscious calculation to a nineteenth-century observer acquainted with whaling, much like the size-to-age ratio a twenty-first-century consumer might make about laptops or mobile phones (smaller equaling newer). In fact, ships and computers share something like a culture of obsolescence. Alexis de Tocqueville asked a sailor why US ships were not built to last, and the sailor allegedly responded that "the art of navigation makes such rapid progress daily that the most beautiful ship would soon become almost useless if its existence were prolonged beyond a few years." Obsolescence in this sailor's view becomes a measure of technological progress. Media critic Jonathan Sterne notes that "this reasoning has been carried forward by an international computing industry."[98]

In short, in voyaging from an old ship from Nantucket, Ishmael focuses his attention on one of the consequences of economic change and transition: on the piling up of old, obsolete things. Melville's attention to the old and obsolescent—Ishmael's explicit refusal of the new and au courant in whaling technology—is in 1851, at peak whale oil—an act of critique. Later, in chapter 24, "The Advocate," Ishmael offers another critique of extractive capitalism's mandate for endless growth by carrying the idea forward to its absurd and troubling logical end. Chapter 24 offers the novel's most straightforward macroeconomic account of the whaling industry. The whaling world described in this chapter is potent and ascendant, but that exultation is checked with Ishmael's derisive humor and monstrous imagery: "I freely assert, that the cosmopolite philosopher cannot, for his life, point out one single peaceful influence, which within the last sixty years has operated more potentially upon the whole broad world, taken in one aggregate, than the high and mighty business of whaling. One way and another, it has begotten events so remarkable in themselves, and so continuously momentous in their sequential issues, that whaling may well be regarded as that Egyptian mother, who bore offspring themselves pregnant from her womb."[99] The pace of reproduction is supernaturally accelerated in the figure of the Egyptian mother bearing pregnant newborns: one generation replaces the last long before its members experience a full lifespan. The exponential productivity of the Egyptian mother's pregnant children characterizes the explosive political events that the whaling industry has set in motion: specifically the colonization of South America, Africa, the Pacific Islands, and the Arctic by European and US imperial powers.

The fecund Egyptian mother figure runs counter to other critiques of extractivism that stress the "no-future paradigm" of extraction: the way that extractive economies steal from future generations and interrupt the reproduction of life.[100] In contrast, the Egyptian mother bearing newborn children, themselves already pregnant, is an image of pure fertility. The figure of fertility, focused on the violent whaling industry, is ironic and even grotesque. The booming American whaling industry, its excessive power as an agent of vast historical change on a global scale: Is its fecundity itself unnatural, perhaps even cancerous? Consider the consequences of the Egyptian mother's pregnant newborns: children become parents, then grandparents, and then great-great-grandparents at an astonishingly fast rate. Not only is this world constantly supplied with new generations; it is also a world populated by great-great-grandparents, older generations increasingly alienated from the world's youngest children. In a world of exponential growth and viral fecundity, the persistence and multiplication of old members is as significant as the rapid introduction of ever-new generations. The rapid pace of change meant that the world was refreshed with new ideas, new machinery, new technology, but more than that, it was a world heaped with the dead and the old: obsolete technologies, depleted resources, extinct species. The pileup of the old, dead, depleted, and obsolescent is a feature of consumer culture, too, which is obsessed with newness and novelty and leads to what Charles Acland called overproduction.[101] The pileup of old, discarded things suggested by the metaphor of the Egyptian mother also forecasts the pileup of toxic e-waste now haunting large parts of the globe.[102] Melville's critique of extraction is somewhat unique in its focus not only on what extraction removes and steals from the future but on what extractive economies leave behind after their inevitable ends.

While technological development emphasizes progress, improvement, and the generation of new forms, the whaling industry of Nantucket with its highly visible obsolete infrastructure demands a different theory. *Moby-Dick* apprehends the drag of the obsolete on technological advancement, and the narrative offers the outlines of a new theory that understands built-in limits on the futurity of capitalism. Capitalist development and technological innovation generate as many old forms as new ones: obsolescence and obsolete forms are both constituent and fatal to growth. Obsolescence is built in.

Moby-Dick anticipates the obsolescence of American whaling and serves as a premature shrine to an industry that had not yet disappeared. *Moby-Dick*'s temporality is complex and queer: it offers a vision of the future, not by dreaming up new forms to come, but by imagining the structures of the present in ruins. *Moby-Dick* narrates energy history in the future past tense, in *futur antérieur*: whaling will have declined. Literary critic Cesare Casarino first described the temporal orientation of *Moby-Dick* as the futur antérieur, the future past, the orientation from which one regards the present moment by thinking of it as it will appear in the future, as history. The strange temporality of the futur antérieur is, according to Casarino, the form that results from the "delirium of the writing of crisis."[103] Casarino all but names "energy" and "energy transition" to describe the social-cultural-economic crisis of the mid-nineteenth century. Even more recently, critics and theorists are resorting to the futur antérieur to describe the predicament of understanding the past and imagining the future of a climate-changed world.[104] As *Moby-Dick* indicates, the futur antérieur is not only a stance for critics but a narrative temporal orientation that can afford creators in any genre the ability to understand the possible future by imagining what is past, or obsolescent, about the present. The artist Robert Smithson wrote, quoting Vladimir Nabokov, that "the future is but the obsolete in reverse."[105] Futur antérieur is the tense and narrative stance for imagining the future obsolete, but one that leaves open the possibility of multiple futures.

Ishmael sees Nantucket, New Bedford, and the booming whaling industry itself as transient episodes in a panorama of rising and—always—declining empires. Ishmael's nostalgia led him to choose a ship slated for doom; the novel, of course, ends when the white whale sinks the claw-footed *Pequod*. This prophetic nostalgia resists triumphant, evolutionary understandings of capitalism. Constant, boundless expansion—"innovation"—is built into the logic of capitalism, but *Moby-Dick* articulates an alternative logic that understands capitalism as a historical process in which obsolescing and obsolete forms accumulate as rapidly as new forms. Unlike the nostalgia that would attract tourists to Nantucket later in the nineteenth century, Melville's premature nostalgia suggests a critique of creative destruction, a prophecy of ruin rather than of explosive growth. *Moby-Dick* became a different book as the whaling industry actually declines, when the quaintness and nostalgia evoked by old-fashioned whaling infrastructure became a widely sought-after experience instead of a perverse prophecy.

PART TWO
Whaling Entertainment

2

The Invention of Quaintness
Nantucket Tourism and the Logics of
Energy and Exhaustion

Published in 1851 at the height of the whaling industry, *Moby-Dick* foresaw its coming decline. The second part, "Whaling Entertainment," explores what happened when Melville's prophetic vision came to pass, and the vanishing whaling industry began to signify leisure and entertainment rather than industry and work. The transformation of whaling culture from industry to entertainment is an example of what media theorist Steven Jackson calls "broken world thinking": a proposition for media scholars to understand that "breakdown, dissolution, and change, rather than innovation, development, or design . . . are the key themes and problems facing new media and technology scholarship today."[1] Broken world thinking applies to energy regimes and infrastructures, too, as this chapter will demonstrate. The work that tourists, renovators, and developers did to refashion whaling infrastructures into sites of entertainment did not follow established practices of repair or maintenance; rather, their work infused new values and associations into Nantucket's ravaged landscape and obsolete infrastructure. The transformation of Nantucket in the last decades of the nineteenth century was a manifestation of broken world thinking that attached new values to old things.

In a sense, many late nineteenth-century tourists were latter-day Ishmaels; when they found themselves "growing grim about the mouth," they headed for the seaside. Many tourists from the Northeast even followed Ishmael's own journey through the whaling landscape of New England, stopping over at New Bedford on the way to Nantucket. Unlike Ishmael, they transferred from modern trains to modern steamships to complete their voyage to the quaint island. Late nineteenth-century New Bedford was industrial: after whaling, the town turned to textile manufacturing and to the project of refitting whaling services for the new fossil-fueled economy. New Bedford whale oil refineries began processing petroleum, and companies that had produced rope and line for the whale fishery were refitting to provide cordage needed for textile machines and, eventually, automobiles.[2] New Bedford wharf-side spaces that had once been piled with whale oil casks were increasingly dominated by coal pockets, huge containers that stored the coal moving in and out of the harbor.

Nantucket, meanwhile, was refitting itself into the seaside tourist retreat par excellence: a quaint setting for a new sort of ostentatious vacationing, away from the noise and smoke of fossil modernity. During the last few decades of the nineteenth century, the island was transformed by a process of radical reinterpretation. In the middle of the nineteenth century, Nantucket was a postindustrial ruin whose economic downfall had been brought on by the decline of its whaling industry. Nantucket was depopulated and increasingly decrepit, its oil warehouses and candle factories stood empty on decaying wharves. Many Nantucketers left the island for work elsewhere, and many who remained were poor. But ruin itself was the source of Nantucket's charm for outside visitors. Visible signs of Nantucket's decay and decline were reinterpreted by the tourist's gaze and the promoter's will as *quaintness*. Local promoters carefully preserved the island's quaint, ruined infrastructure, and actively built a tourist infrastructure to support visitors who wanted to experience it: developers built cottage cities, resort hotels, a real estate industry, and public transportation in the form of a little railroad connecting the island's towns.[3] In the last part of the nineteenth century, Nantucket transformed from an impoverished backwater to a fashionable resort whose obsolete infrastructures had been carefully preserved and prettified in the service of augmenting quaint charm for summer visitors. Today, in the twenty-first century, Nantucket and New Bedford are studies in contrast: well-heeled Nantucket thrives on preppy tourism and eye-popping real estate sales while New Bedford centers on a large and imperiled fishing industry.

The quaintness tourism that sustained Nantucket in the late nineteenth century was predicated on two different types of exhaustion. Quaintness is, on the one hand, the aestheticization of economic exhaustion and postindustrial decline. In another sense, Nantucket became a place for exhausted and mostly white middle-class workers from eastern cities to restore their bodies and minds for future work. It is surprising that exhausted workers found in Nantucket the *antidote* for their own exhaustion. According to what Cara Daggett calls the "energetic model of work morality," in which productivity was deemed morally good and waste evil, a deindustrialized town like Nantucket should have been repulsive: a place of waste, ruin, and lost opportunity. Nonetheless, countless testaments from late nineteenth-century tourists, and the enduring importance of quaintness tourism to the present day, indicate otherwise.

Quaintness tourism on Nantucket is a site where transitions in infrastructure and energy were on view for all to see. Old infrastructures related to the bygone whaling industry formed the basis of tourists' attraction to Nantucket. At the same time, the new and expanding infrastructures of fossil modernity—new rail and steamship lines, and hotels built and operated by those railroad companies—made that type of sightseeing possible. During the early years of the development of tourism in Nantucket in the 1870s and 1880s, the Old Colony Railroad held a near monopoly on rail and steamship travel, providing rail and steamship access to southern New England and the islands of Martha's Vineyard and Nantucket. The Old Colony Railroad opened in 1845 with small lines between Boston and Plymouth, and Fall River and Myricks, Massachusetts. The lines ran throughout southeastern Massachusetts and eventually, through mergers and partnerships, all along the New England coast and with carefully coordinated connections to rail and steamship lines in Boston. Some steamship lines even owned and operated hotels, providing the means of travel and the destination all in one company. An example of this nexus is the Nantucket Steam Boat Company, which not only ran the steamboats to Nantucket but also owned one of the popular early Nantucket resorts, the Ocean House Hotel.[4] In the last three decades of the nineteenth century, the Old Colony Railroad in large part built coastal New England tourism and drove and determined, in particular, the tourist economy of Nantucket.

On Nantucket, the hard infrastructures of the Old Colony trains and ships met cultural infrastructures: the evanescent aesthetic of quaintness and the affective experience of exhaustion. Most scholarship in the energy humanities has focused on the hard infrastructures of energy extraction, production, and consumption.[5] But it is important to read such aesthetic experiences as quaintness and such embodied experiences as exhaustion as part of energy history in order to show the connections between energy culture and the vicissitudes of individual emotional experience. Several discourses of energy come together in New England tourism at the end of the nineteenth century: the rise and fall of resource regimes, the development of massive new transportation infrastructures, and the individual experiences of nervousness, rest, and relaxation. The history of Nantucket tourism helps us better understand exhaustion and even aesthetics such as quaintness as parts of energy culture.

This chapter explores material and cultural conditions that made quaintness tourism possible in the late nineteenth-century Northeast: the environmental, economic, and infrastructural transformations that accompanied the transition from an organic to a fossil-fueled energy regime. Mass tourism in New England is a phenomenon not only of fossil modernity but of the obsolescing resource regimes that fossil modernity eclipsed. At the center of this chapter is a constellation of travel narratives and tourism brochures from Nantucket and southern New England, which testify to the aspirations and experiences of tourists from the mainland and island hosts: a varied body consisting both of literary narratives and promotional and prescriptive material that I group together under the name tourist literature. Tourism changed the meaning of industrial infrastructures in New England whaling ports, turning the tools of a brutal extractive industry into objects that invest a visit to the seaside with pleasant melancholy, quirky charm, and visible history.

QUAINTNESS AND POSTINDUSTRIAL TOURISM

The urge to visit ruins was not new to the late nineteenth-century United States. For more than a century, wealthy Europeans on the Grand Tour contemplated and sighed over the ruins of ancient Rome, Greece, and Egypt, finding among toppled columns and ancient pyramids lessons in hubris and a gratifying melancholy that supposedly opened up the minds of young travelers to various truths about mortality, the ravages of time, and the folly of the human drive for grandeur and personal fame. The urge continues, recently manifesting in the fad of "ruin pornography": aestheticized photos, films, or prose about abandoned factories, wrecked public buildings, or deteriorating monuments. Ruin porn highlights the contrast between the clean geometric lines of a given structure's original architecture, or the finely-wrought details of its decoration, and the chaotic textures of peeling paint and piled debris. Detroit, Michigan is an American capital of ruin porn and a related phenomenon: tourism carried out by so-called urban explorers who bolt-cut fences and padlocks to enter abandoned buildings and take the same overcast pictures of decrepitude as everyone else.[6] Favorite sites of ruin pornographers and tourists include such places as the closed auto factory Fisher Body Plant 21, a luscious movie palace from 1921 since turned into a parking garage, and the towering monumental train depot called Michigan Central Station: all poignant symbols of fossil-fueled transportation culture. Detroit ruin porn rarely, if ever, accounts for actual Detroit residents: for people impoverished by the economic exploitations prettily transformed through ruin porn. Ruin porn specifically excludes and erases Detroit's Black communities and culture, which is especially notable considering Detroit is a majority-Black city. As John Patrick Leary, one of the shrewdest chroniclers of Detroit ruin porn, puts it: "So much ruin photography and ruin film aestheticizes poverty without inquiring of its origins, dramatizes spaces but never seeks out the people that inhabit and transform them, and romanticizes isolated acts of resistance without acknowledging the massive political and social forces aligned against the real transformation, and not just stubborn survival, of the city."[7]

Despite differences in place and time, ruin porn–based tourism in Detroit in the early decades of the twenty-first century is an analogue to late nineteenth-century quaintness tourism in Nantucket.[8] Detroit ruin porn is driven by some of the same logics of energy and exhaustion as Nantucket quaintness tourism: ruin porn offers privileged, out-of-town visitors the opportunity to confront exhaustion on a monumental scale from the safe distance of privilege, where disaster titillates rather than threatens. It seems clearer to critics of twenty-first-century Detroit ruin porn that industrial ruins also allow spectators to imagine a threatening future. Detroit has long been seen, as Leary puts it, as a "bellwether of each major urban crisis since World War II," with early twenty-first-century Detroit ruin porn forecasting the "looming jobless future, or more precisely, our worst fears about the future" from within the uncertainty of the Great Recession in 2009.[9] Such was the critical perspective on Nantucket in *Moby-Dick* that I explored in chapter 1: from the perspective of Nantucket, Ishmael was able to imagine the then-prosperous whaling industry of the mid-nineteenth century in ruins, to see the end of the extractive resource regime.

For the nineteenth-century quaintness tourists and tourism promoters on Nantucket, the threats of inevitable exhaustion or a future in ruins are sublimated into an enthusiastic appreciation for quaintness and into anxiety about other types of exhaustion: the exhaustion of body and mind in the individual modern office worker. The tourist gaze on late nineteenth-century Nantucket focused not on the threatening future but on a safely distant past. Nantucket quaintness tourism and Detroit ruin porn bookend fossil modernity: Nantucket quaintness chronicled the uneven energy transition of the late nineteenth century, and Detroit ruin porn the ongoing and uneven energy transitions of the twenty-first century. It is no coincidence that Nantucket quaintness grew on the abandoned industrial spaces of wharves and wooden sailing ships, on such obsolete implements as harpoons and barrels, or on the exhausted natural resource of the whale's body, just as it is unsurprising that Detroit ruin porn focuses on spaces of fossil modernity: car factories, parking garages, and train stations.

FOSSIL-FUELED TOURISM

The coming of fossil fuels was received ambivalently in the nineteenth century on Nantucket, whose wealth was built from whaling fortunes, and whose city streets were lit with whale oil brought home by islanders. Likely owing to local pride, Nantucket was slower than many other regional towns and cities to relinquish whale oil lamps and install gas lights for lighting in public spaces. According to Edward K. Godfrey's guide to the island in 1882, "Nantucket, which had for two centuries given light to the world, at last accepted herself the inevitable, and in 1854 gas was lighted for the first time in the town."[10]

But much as local Nantucket might have mourned the transition from whale oil to natural gas in its streetlights, the island's tourist industry and its postwhaling prosperity were made possible at every level by coal, petroleum, and natural gas. Between 1865 and 1890, the number of railroad miles in the United States increased fourfold, and fossil-fueled trains brought tourists to Nantucket.[11] On the most practical level, the railroads increased access to new coastal resorts and made it faster and easier than ever for passengers to visit even once-remote spots along the New England coast. But the railroads and steamship lines did much more to create the experience of New England tourism than provide logistical access: rail and steam lines financed the construction of hotels, making resort destinations more viable and attractive to tourists. And even more intriguingly, the railroads produced a huge body of tourist literature in many genres, from timetables to guidebooks to fictional narratives, which scripted the experience of New England tourism from transportation logistics to aesthetic and affective response.

The fossil-fueled infrastructure of rail and steam that networked New England was spectacularly visible: an attraction in itself even to the casual tourist who picked up a brochure to plan a vacation.[12] The prolific body of print literature and ephemera produced by the railroad—guidebooks, narratives, and functional items such as schedules, tickets, and ticket envelopes—are part of that infrastructural system. It is neither new nor surprising to learn that the railroads and steamship lines would work to make the tourist destinations they served and invented more attractive through print advertising. Indeed, the production and reproduction of stories and images of tourist sites are some of the foundational processes of tourism everywhere: what tourism historian Richard H. Gassan calls the "cultural infrastructure" of tourism.[13] In coining that term, Gassan anticipated a gesture now common in the new energy and infrastructure studies, which promotes a capacious understanding of infrastructure assemblages and careful attention to the cultural forces and material substances that combine in the construction of fossil fuel infrastructures.[14] The print culture that was generated by and about the steam and rail lines have to be understood as parts of the infrastructural assemblage as much as the steamships and railroad engines themselves. In considering print culture a constituent piece of the infrastructural system of emerging New England tourism, there is no meaningful boundary to be drawn between the utilitarian printed ticket and the beautifully crafted work of local color or literary regionalism. The mass construction and adoption of fossil fuels in the nineteenth century is not restricted to material technologies but extends also to the way that various cultural forms and individual affects are embedded in infrastructural development. Through late nineteenth-century tourism, literature itself gets bound ever more tightly around fossil fuels. And the purpose of so much of the cultural and literary infrastructure of Nantucket tourism in the late nineteenth century was to give prospective travelers a reason to go there and not to another seaside resort. Nantucket tourist literature could not rely on longstanding myths about classical ruins but had to tell the story of a world-famous enterprise that had only just ended. By spinning the signs of Nantucket's recent history into emblems of quaintness, tourist literature worked alongside real estate developers and local promoters to build the island's tourist infrastructure.

Some of the island's most interesting cultural infrastructure was produced directly by the railroads and steamship lines. The Old Colony Railroad's maps and timetables were enticingly elaborate, beautifully engraved, and adorned with historic timelines and scenes of local interest.[15] Beginning in the 1870s, the Old Colony Line also published a series of guidebooks for "the sportsman and tourist, and for all in search of rest and recreation."[16] The guidebooks are heavily illustrated, paperbound pamphlets, with timetables printed on the front and back inside covers and an extensive directory of hotels convenient to the Old Colony Lines. Pasted into most of the guidebooks is a foldout map of "The Old Colony Railroad and Connections." Maps in different editions of the guidebook produced throughout the 1870s and 1880s index the growth of the rail lines; over the years, the names of the rail road shift, as Old Colony merges with the Fall River Line and the Boston & Nantasket Steamboat lines.[17] Through the end of the nineteenth century, the foldout maps get larger and denser with new rail and steamship lines, and the guides get longer, more intricate, and more varied in form and genre.

Most Old Colony guidebooks are written from a first-person or limited third-person point of view. The voice of the unnamed narrator is generally pleasant and helpful, as in this passage that opens *The Popular Resorts and Fashionable Watering Places of Southeastern Massachusetts and Newport, R.I.*, published in 1878: "While passing a few weeks['] vacation along the shores and through the forests in the eastern section of the State, the writer of these pages was surprised and delighted with the great natural beauty of the country."[18] In most guides, the narrator's identity is little developed beyond his logistical needs. For this narrator, for example, "the requirements of business were such as to necessitate his return to the city every few days. Herein he realized the convenience of access afforded on the numerous lines and branches of the railroad whose iron pathways compass this vast district."[19] The assumed narrator of the guide is a heavy user of the railroad, as he darts between his summer vacation site and the workplace back in the city: an ideal frequent railroad customer and a person whose complex logistical needs surely console the average vacationer whose transport needs are less complex.

The railroads also published more ambitious and literary first-person (presumably fictional) narratives in the form of records of a specific character's travels. The narrative *Old Places and New People; or, Our Pilgrimage and What We Saw* (1881) tells the story of Silas Winfall, a middle-class New York magazine editor, who travels with his family to visit newly discovered extended family who all live along the Old Colony Railroad. Another fictionalized guidebook, *The Tip End of Yankee Land*, published in 1885 by the Fall River and Newport Lines and the Old Colony Railroad, follows the madcap adventures and flirtations of a young, wealthy, and bored Boston bachelor named Augustus Oliver as he traveled with a group of friends to Cape Cod, Martha's Vineyard, and Nantucket. The adventures of Silas Winfall and Augustus Oliver are aided at every step by the sponsoring railroads, whose convenient schedules and comfortable accommodations are described in extensive and practical detail.[20] These narratives are written in a light and charming style, and their protagonists, at least, are more fully developed as characters. Their variety and difference—the middle-class, middle-aged Winfall and the dashing young Oliver—speak to an advertiser's interest in both expanding and segmenting its consumer base.

Unsurprisingly, the guidebooks and narratives testify in glowing terms to the comfort, convenience, and luxury of the specific rail and steamship lines that they advertise. The infrastructure and conveyances are spectacular sites in themselves, especially in the early guidebooks from the 1870s. Around half of the *Popular Resorts and Watering Places of the Old Colony*

Line (1877) is dedicated to extolling the technological marvels of travel in the age of coal, steam, and oil and the specific efficiency and luxury of the Old Colony rail and steamship lines. The guide details everything from the technical specifications of the boilers to the "large and sumptuous" staterooms and the "ample tables" in the ship's parlors.[21] But the narrative also extols the virtues of arcane transport logistics, the complex labor of planning routes and coordinating schedules so that the Old Colony Line might meet commodiously with other regional transports:

> The perfect system which marks the completeness of this railway is not the work of an hour: it demanded years of toil, and an outlay gigantic in its character. The growth of Boston shows the business value of such a roadway. Hundreds of towns and villages have been called into existence. Countless manufactories have sprung up in the valleys. Sterile farms, that scarcely afforded for man or beast, teem with a busy population, making the poor farmer a rich man. Men who dwelt in the lanes and narrow tenements of the city now breathe the healthy air of the country, and get to business or labor cheaply and on time. A railroad in a town is like a mine of gold; and land, produce, and labor feel its magic power. The snorting of the iron horse indicates prosperity.[22]

The railroad itself is credited with enriching the region and shaping the prosperity and even the bodily health of individual people. More novel than the cliché "iron horse" metaphor, the comparison of a railroad to a "mine of gold" evokes the two-faced nature of extractive industry, the looming threat of the resource curse as well as the wealth that new transportation logistics might stimulate. The Old Colony guidebooks and narratives self-consciously situate the Old Colony Line within other infrastructures of manufacturing, passenger, merchant, and extractive shipping. Infrastructure became part of the spectacle.

In addition to teaching passengers about rail and steam logistics from the points of view of technology and the regional economy, the guides prepare passengers for their experience on board. Some of the print material is explicitly pedagogical. An extensively adorned small envelope whose purpose was likely to hold tickets fulfills that function. A portrait of a steamer of the Fall River Line near shore at "The Dawn of Day on Narragansett Bay" covers the front of the envelope and hints at the tranquility of a summer vacation, while the back of the envelope is crammed with all of the intricate information needed to get there. The triangular envelope flap is printed with a list of instructions for passengers, inciting them to "PLEASE REMEMBER":

THAT TICKET should be removed from Envelope before it is handed to Collector at the gangway. . . .

DO NOT LEAVE Personal Effects in Stateroom or Berth. Look around carefully to see if you have left anything.

SEATS IN PARLOR CARS from Fall River may be secured at Purser's office.

EAST-BOUND PASSENGERS are called forty minutes previous to departure of first Boston Express train, unless contrary notice is given at Purser's office. . . .

THE ELECTRIC LIGHTS in Staterooms are turned on and off the same as gas. To call a servant, press the electric button a second.[23]

The back of the envelope represents the two-part infrastructural system of steamship and railroad that brings passengers to fashionable watering places. These commands are printed discreetly, on the inside of the envelope flap, perhaps so as not to embarrass a novice passenger just learning the ropes. The envelope is something between a prescriptive text and what cultural historian Robin Bernstein calls a "scriptive thing, an item of material culture that prompts meaningful bodily behaviors."[24] As these representative instructions from the ticket envelope indicate, passenger travel on the steamships of New England was tightly choreographed, so that passengers could slot themselves into the matrix of steamship-to-railroad connection that made systems such as the Fall River Line so "convenient." Passengers are instructed in how to move through the connection from the moment that they present the ticket for boarding ("ticket should be removed from envelope") until the moment they leave their staterooms. The Purser's office on board ship is also a railroad ticket office, where parlor seats from Fall River might be booked. Toward the end of the century, railroad tourist guides devote less space to transport logistics and more space to the destinations, reflecting, perhaps, the growing familiarity of the vacationing public with transport logistics and comportment.

THE AESTHETIC CATEGORY OF QUAINTNESS

Quaintness, Nantucket tourism's most celebrated aesthetic, was the result of a pleasing disjunction between that "modern" world of fossil-fueled infrastructure and the on-island world that seemed untouched by it. But travelers wanting to escape from fossil modernity came to the island by means of smoke and steam and enjoyed its quaint attractions thanks to thoroughly modern practices of real estate and infrastructure development. The experience of quaintness was not possible without direct experience of its imagined alternative, modernity.

The journey by train and steamship from New York or Boston to Nantucket prepared tourists for quaintness by offering glimpses of the stark contrast between the quaint and the fossil-fueled modern. The promotional literature prepared tourists for a smooth, modern experience on new railroad cars and luxurious steamships, but the infrastructural systems that tourists encountered in coastal New England were messy and complex. In addition to navigating railroads and steamships, tourists still moved through and alongside older, crumbling infrastructures. A photograph in the collection of the New Bedford Whaling Museum portrays the collision of old and new infrastructures at the landing site of the steamship *City of Taunton* in the 1890s.[25] In the background is the *City of Taunton*, with its straight lines, geometric windows, and tall smokestack, and in front of the ship, a queue of horse-drawn carts and passengers process down the wharf. Moored in the foreground along the passengers' wharfside procession are two old square-rigged whaling ships, broad-beamed and with their tall masts askew, compared with the straight, vertical line of the steamship's smokestack. Crammed untidily between the whaling ships and the passengers are vast barrels for whale oil stacked amid piled fishing nets, barrel staves, and other nautical debris. The ships at the wharf, the empty barrels, and the untidy debris speak to the decline of whaling, and the proximity between the thriving steamship business and the visibly obsolescing whaling industry makes the contrast all the starker.

For tourists to New England's burgeoning coastal resorts, New Bedford became a place to stop only while transiting from the railroad to the steamship lines. An early tourist's account of traveling to Martha's Vineyard, previously cited in this book's introduction, noted a stopover in New Bedford and the chance to observe the "grim whalers idling in the sun." This tourist imagined the whaling ships as "frowning hulks," envious of working vessels: "I fancied that they looked more melancholy, and felt a twinge of envy, as a noble ship shot out from Edgartown across our steamer's bow, her every sail set."[26] Later, guidebooks cued tourists transiting in New Bedford to pay attention to the old ships but not to linger too long in the declining city. *A Guide to Martha's Vineyard and Nantucket*, published in 1878, includes a short description of New Bedford but places it in the "Routes of Travel" section rather than according it destination status: "Although New Bedford has lost its old-time prestige as a seaport, it still retains many of its ancient ships and much of the paraphernalia appertaining of the whale-fishery.... The large warehouses, standing like grim sentinels along the water-front, in their somberness are suggestions of what the city once was, rather than indicative of the presence of modern prosperity."[27] This writer imagines the whaling ships as paralyzed, sleeping

Passengers from the steamship *City of Taunton* disembark on Steamboat Wharf in New Bedford alongside aging wooden whaling vessels, likely in the 1890s. Courtesy of the New Bedford Whaling Museum.

hulks, and to see the steamship that bears the narrator on the first leg of the journey from New York as a being in perpetual motion: "The steamer at the pier hisses, splutters, and groans at you as you hurry down with your carpet bag."²⁸ The sight of the still, obsolescing ships was by no means repulsive: the ships stirred the imagination, nearly coming alive through almost-heard growls. Dilapidation was appealing but, in New Bedford, not exactly quaint.²⁹

Nantucket, by contrast, was quaint through and through. In New Bedford, tourists experienced the clash of old and new: modern coal-powered steamships docked alongside decaying wooden whaling ships. Nantucket, the other great US whaling capital, had experienced the loss of its whaling industry much earlier and had not, for reasons of its offshore geography, become a bustling transit point on rail and steamship lines. Compared with New Bedford, the obsolescence of whaling in Nantucket was well advanced by the time tourists began to arrive in large numbers in the 1870s.

By the 1880s, Nantucket was the capital of quaintness. "Quaint" was an aesthetic, an attachment, a feeling that attached to all of New England, but Nantucket was deemed by many of its visitors to be the quaintest place of all. Nantucket quaintness is capacious and ineffable, attaching to descriptions of architectural style, people, literature, history, and general atmosphere. Edward Godfrey's aptly named encyclopedic guidebook on *The Island of Nantucket: What It Was and What It Is* addresses quaintness under the heading "ARCHITECTURE": "Newspaper correspondents, and in fact visitors generally, describe the town as a quaint old place. The town is old, as old things go in this country, and there is a very strong smack of quaintness about those houses which Burdette so happily describes as being shingled, shangled, shongled, and shungled."[30] By the end of the nineteenth century, "quaint Nantucket" was already a well-worn cliché that visitors could not invoke without self-consciousness. A letter from a visitor reprinted as a testimonial for the Sea Cliff Inn laments the overuse of the term: "Well, here I am, safe and sound, in quaint old Nantucket. It is too bad that adjective 'quaint' has been so promiscuously used, for I am very sure the copyright of it should belong solely to Nantucket."[31]

As an aesthetic category, quaintness is not nearly so serious as the beautiful or the sublime, not as widespread in its possibilities as the picturesque.[32] The experience of quaintness is clichéd to the point of embarrassment for those who experience it. I propose a more substantial examination of quaintness, nevertheless, following the example of Sianne Ngai, who forged the theory of "minor aesthetic categories" in her study of the cute, zany, and interesting. Ngai's study shows that these aesthetic categories are—by virtue of their very triviality and ubiquity—"the ones in our current repertoire best suited for grasping how aesthetic experience has been transformed by the hypercommodified, information-saturated, performance-driven conditions of late capitalism."[33] But it is precisely because minor aesthetic categories are more accessible, available, everyday aesthetic categories that, according to Ngai, their study expands our understanding of aesthetic experience.

There is a strong affinity between the aesthetic categories of "cute" and "quaint." For Ngai, the aesthetic category of cuteness arises in the relationship between a consumer and a commodity. Cuteness is "a kind of commodity fetishism, but with an extra twist": the diminutive, powerless, cute commodity object invites its own purchase as a kind of motherly custodianship by a consumer, who is manipulated by cuteness to buy it.[34] Ngai traces the etymology of "cute" to the mid-nineteenth-century United States, where it is at first a shortened version of "acute" and an unequivocally positive aesthetic.[35] The aesthetic slowly transmogrifies to the equivocal aesthetic experience in the twentieth century, tracking the expansion of consumerism in the United States. Cuteness thus helps us understand the particular social and historical formation known as postwar consumerism.

As I show in this chapter, quaint things and people, like cute things and people, are judged powerless by beholding subjects; the appreciation of both cute and quaint is tinged with cruelty and violence, as if the beholding subject wanted to enforce the cute or quaint thing's powerlessness. Essayist Daniel Harris has offered a series of reflections on the quaint—along with the cute, hungry, and romantic—as "aesthetics of consumerism." Particularly relevant here is Harris's reflection on how quaintness reflects a consumer's patronizing attitude toward people in the past: patronizing dismissal is one of the forms of the cruelty that beholding subjects exercise on the quaint object.[36] The effect of quaintness, like that of cuteness, sometimes results in changes to language; cuteness is registered in the reduction of adult speech to babytalk, an effect mimicked by the sing-song description of Nantucket's "shingled, shangled, shongled, and shungled" houses quoted above.

But quaintness is not merely a variant of cuteness. Quaintness is an aesthetic category born, not, like Ngai's cuteness, in cultures of mass consumption in the twentieth century, but in the cultures of energy, infrastructure, and technology kicked off by the energy transitions of the nineteenth century. A Google Ngram, charting the usage of the term "quaint" in books held by the Google library and published between 1800 and 2000 falls roughly along a bell curve that peaks just before the turn of the twentieth century, in exactly the period during which New England regional tourism peaked. Quaintness helps describe the social processes that accompany environmental and technological changes. Specifically, quaintness describes a particular relationship to work and leisure as they are mediated by energy transition. Following Ngai, I might define Nantucket quaintness as a way of aestheticizing power in decline—or, perhaps more to the point, as a way of aestheticizing exhausted resources, industries, and people. The strange characters tourists encounter on Nantucket—quaint old sea captains and an anachronistic town crier—were figures of power in an earlier age, but their power and energy have been stripped by age and the movement of US industry away from the coast.

Quaintness is a special kind of relationship to the past. Daniel Harris calls quaintness a fundamentally "patronizing" judgement, but Grace Lavery is inclined to be more generous in her appraisal of the quaint. Lavery traces the emergence of quaintness across roughly the same chronology followed here—from the mid-nineteenth century through the early twentieth—but in a different archive: Anglophone Victorian writing about Japan, a place that was obsessively described as "exquisite" and, like Nantucket, "quaint." According to Lavery, quaint things do not become historical in predictable or expected ways. "[Quaintness] referred to an oblique slippery relation to history, a distinctive mode of passing into the past. Nothing is quaint from the get-go; an object, text, body, or event acquires the quality of quaintness as it becomes historical—or, more precisely, as it *fails* to become historical."[37] Lavery recuperates quaintness and even makes quaintness the aim of her own critical and historiographical method, "which seeks to activate quaint attachments in order to develop a richer engagement with obsolete aesthetic categories than traditional historicism generally accesses."[38] I follow Lavery's example in bringing the "obsolete aesthetic category" of quaintness to bear on my historical inquiry into energy transition and its effects on coastal leisure, although I do not think that the Nantucket quaintness discourse fulfills the promise of quaintness to open up space for nonnormative behavior; as I discuss below, Nantucket quaintness upholds whiteness and wealth.

Quaintness was an aesthetic experience made possible by the knowledge that things were different before on Nantucket and elsewhere. Nantucket's quaintness was marked as the difference between Nantucket and the modern world ashore and, at the same time, as the difference between Nantucket's impoverished present and its prosperous past. Thus Nantucket's quaintness was an effect created in part by those things that have disappeared: namely, the whaling industry and the prosperity it brought to the island. By the late nineteenth century, the whaling industry of Nantucket existed mostly in the memory of locals and the imagination of tourists. The whaling industry's demise had a lasting emotional effect on the island's population, as described in the account of Nantucket's whaling history written by Alexander Starbuck in 1882 and published in Godfrey's guidebook:

> At home, when peace reigned, the people were all busy, happy, and prosperous, the warehouses were crowded with goods, and the streets thronged with teams and foot passengers. At the wharves lay a large fleet of vessels taking in or discharging cargoes or refitting for new voyages. The cheery din of the cooper's hammers and the ring of the blacksmith's anvils resounded on all sides, the sail lofts, the shops of the riggers, and the "walks" of the rope-makers were occupied by the multitudes that the demands of the shipping gave employment to. In a thousand ways the activities of a prosperous business showed

themselves. But all this is now changed. The ships long ago sailed on their last voyages from Nantucket.

Not an ocean on the face of the globe but holds in its embrace the shattered remains of a portion of her fleet, while the surviving portion hails from other ports. The tools of the mechanic are silent, and the bustle of traffic no longer crowds the streets. The wharves are deserted, decaying, or decayed, and the warehouses have long been vacant and closed. To a native of Nantucket, it is a sad sight to thus see "Ichabod" written on her desolate places; to look upon the ruined wharves and storehouses, and to see even the "toilers of the sea" themselves look old and weather-beaten; to see them rapidly nearing that port in which the anchor will be cast never to be weighed again.[39]

The passage of the whaling business from Nantucket is rendered in Starbuck's description as an evacuation of sights and sounds: tools are "silent," "traffic no longer crowds," "wharves are deserted," warehouses are "vacant," and even the old whalemen are attenuated and lifeless. Part of Nantucket's quaintness owes to its quietness and the perceptible absence of bygone industry.

Death, ghosts, and haunting were the ruling metaphors for the passage of the whaling industry from the island and the visible signs of its decline. One account from 1868 in *Lippincott's* compares Nantucket's deathly appearance to that of quaint European towns: "Nantucket now has a 'body-o-death' appearance such as few New England towns possess. The houses stand around in faded gentility style—the inhabitants have a dreamy look, as though they live in the memories of the past."[40] For tourist A. Judd Northrup, whose account of a summer spent on Nantucket helped popularize rustic cottage vacations on the island, the memory of whaling haunts the island like a ghost: "The wharves are ample to receive the oily freights of many whalers, as in the good old days, if only their ghosts would rehabilitate themselves in oaken hulls and spread again the many-sheeted canvas; but they are nearly all vacant now."[41] Northrup's account of empty warehouses aligns closely with Starbuck's; the wharves had become spaces of eerie sensory deprivation, in contrast to what they once were and were meant to be. But unlike Starbuck, Northrup is less reverent of that old history. Northrup and his family also visit the Nantucket Athenaeum, where some of the old whaling implements and artifacts are exhibited, and he is intrigued but ultimately put off by the fetishization of these old objects. "We went industriously about the town, visiting various resorts of special interest;—and first, the Athenaeum, which contains the Library and Museum, where they serve you up whales' jaws, teeth and so on, and harpoons,—indeed every interesting thing appertaining to whaling enterprises, except a wreck or a man overboard,—besides the usual dusty and musty antiquities that give a ghostly sanctity to museums, the dead-houses of the past."[42] Northrup and his family pay respect to the artifacts in the library and museum "industriously," which is to say dutifully and grudgingly. Northrup regards the museum as one of the "dead-houses of the past" and the preservation of whaling history there as a type of necrophilia. During his time on the island, Northrup developed a fine-tuned taxonomy of quaintness. We know from the bulk of Northrup's narrative describing his family's idyllic summer that Northrup deeply values the quaintness of the little village he inhabited. But for Northrup, quaint objects and people incite pleasure when they stand in place, unwitting relics. The preservation and self-conscious exhibition of old artifacts in a museum, on the other hand, does not give Northrup the same experience of quaintness. Quaintness is a fleeting and a fugitive aesthetic.

The tone in which Nantucket quaintness is invoked varies across texts, ranging from somber and sentimental to light-hearted and dismissive. A fictional narrative published by the Old Colony Line exemplifies the latter. Published in 1881, *Old Places and New People; or, Our Pilgrimage and What We Saw* narrates the story of the "Winfall" family who use a newly discovered family tree to visit distant relatives settled along the Old Colony Line. A pitchman for the railroad and steamships, the narrator carries on obsequiously about the "splendid floating hotels" of the steamships and the "swift express" rail lines. But at the same time Winfall carefully describes historical infrastructure, too. He is taken with the decaying whaling infrastructure on Nantucket, and he prophecies that its ruins will become emblems of a new prosperity as Nantucket becomes an ever more fashionable summer watering hole: "Let the old ships rot where they are stranded! Let the swift and keen harpoon rust upon its hooks! 'Tis bric-a-brac now, and I would like to have it for my library. There are other days for 'old Naintuck.' When the stumpy mariner shouts, 'There she blows!' be sure he has heard the steamer's whistle, and more tourists are coming, and times will be 'flush' again."[43] Winfall is exuberant in his cynicism: he imagines the preservation of whaling implements not in the "dead-house" of a museum but in the confines of a private library where harpoons become bric-a-brac for the man who vacations on Nantucket and carries its rusting infrastructure back home to serve as pleasant curiosities and souvenirs of a relaxing island holiday. The process that Winfall describes—the relocation of old whalecraft from their place on abandoned wharves to the homes of collectors—is in fact the process by which Nantucket was transformed from a postindustrial port town into a pretty resort that had fully commodified its quaintness. Unlike those accounts of quaint Nantucket that are firmly grounded in the memory or imagination of its past and the visible (if picturesque) signs of its future decline, Winfall welcomes the future. Specifically, as Old Colony spokesman, he welcomes the fossil-fueled future encapsulated in the metonymic sound of the steamship's whistle, which overwrites Nantucket whaling history with a call to its tourism future. "There she blows!" signals the advent of a new extractive resource: tourist dollars.

QUAINT PEOPLE

People on Nantucket were quaint, too: like decaying docks and warehouses, they appeared to visitors as unlikely survivors from the past. The opportunity to become a "tourist-turned-anthropologist," as Dona Brown put it, was one of the special features of Nantucket tourism. The tourist accounts of Nantucket are filled with descriptions of the island's old residents, called "eccentrics" and "characters."[44] Perhaps the most famous Nantucket eccentric is the town crier, who circulated in town and declaimed news and announcements. He appears in a tourist guide to "popular resorts and fashionable watering places" in 1878 as the living epitome of Nantucket's quaintness: "The town is a quaint old place, quite unlike any other on the coast. Auctions, meetings, lectures, and even arrivals are announced by the town crier, who of himself is a relic of ancient days. He is almost too omnipresent, greeting one first with his fish horn—toot, toot, toot! and then with his bell—ding, ding, ding!"[45] The description of the town crier serves to elaborate the term "quaintness": the town crier is the chief example of how the town is "a quaint old place," himself a "relic" and a representative of an old way of dispersing news, long outdated by the newspapers and magazines and, especially in this era, the telegraph. Accounts of the town crier focus on the loud racket he makes in town, through dismissive terms such as "bawling" and onomatopoetic accounts of the noises he makes, "toot, toot, toot!" and "ding, ding, ding!" Quaintness veers close to cuteness in this childish language and, as in Ngai's account of cuteness, incites "the radical breakdown of traditional familial forms like the generation," as the town crier flickers between his status as an ancient relic and a babbling toddler.[46]

A small paper hand-bound book called *Nantucket Characters: Indelible Photographs*, produced by Henry Sherman Wyer for the tourist trade in the late nineteenth century, offers portraits of Nantucket's quaint inhabitants. The small book contains twelve images, reproductions of ten photographs and two engravings, each with a title and occasionally a caption.

The famous town crier has pride of place in the book's first portrait. Framed by an open door of a shingled building, the town crier in his rumpled suit and misshapen hat faces the camera, a bundle of papers in one arm and the famous "fish-horn" that he uses to augment his voice in the other. The image is captioned with a representative announcement: "Now, there's been a fearful storm out west! Ar-r-ripping fire in New York! Horrible murder in Boston! Auction this morning—Corn Beef! And a Grand Ball this evening at the rink, and that's the news of to-day!"[47] According to this caption, the town crier reports events of incommensurate scale and importance with each other, a fire in New York in the same breath as the announcement of a corned beef auction. The effect of the caption, in combination with the serious expression on the bedraggled man's face, is to make the town crier silly in his overseriousness. The portrait is at once tender and condescending.

Other portraits in the book feature a ferociously bearded "Old Blacksmith" at his forge; an old woman wearing a lace bonnet and sitting in her rocking chair, captioned "Ninety-Seven"; "A Member of the Society of Friends," an old man in classic Quaker dress of a long black coat and shiny top hat, seated in classic portraiture pose but on a simple wooden chair and with his hand on a modest wooden cane; and a young girl in sailor dress and an oversized straw hat holding a small kitten, her portrait titled "The Pet." Nearly all of the portraits in *Nantucket Characters* are either very old people or young children. Another picture of another pet depicts "The Captain and His Pet," and like the picture of the girl and her kitten, this portrait shows its protagonist literally clinging to a prized possession. Wyer's photograph shows an old man in a baggy suit and hat standing next to a wooden ship's figurehead of a woman in a flowing gown, displaced from her ship and standing in the grassy yard of a shingled barn. A ship's nameplate, "Shanunga," hangs above the barn door. The old man's direct stare and smile are of a grandiosity suiting the beautiful figurehead and the elegant lettering of the ship's nameplate, but at odds with his disheveled suit and the humble, grassy yard in which his ship's adornments rest. The contrast between the captain's overseriousness and the modesty of his setting produces the same effect of appreciative tenderness and pitying condescension that characterize almost all of these photographs. *Nantucket Characters* at its cruelest seems to suggest through the provocative arrangement of photographs and captions that the outdated subjects of these photographs should be embarrassed by their shabbiness, by their claims to importance and grandeur, and even by their survival in a world that needs them only to serve the aesthetic experience of quaintness. From the elevated point of view of the elite viewer—cast into the role of modern

The portraits of young and old Nantucketers in this tourist souvenir book exemplify "quaintness," an aesthetic that romanticizes exhaustion. Henry S. Wyer, *The Captain and His Pet*, photograph in *Nantucket Characters*, 1892. Courtesy of the Nantucket Historical Association, MS397.

by contrast with the people in the portraits—to be quaint would be embarrassing, but to enjoy the quaint is sophisticated. The unembarrassed, guileless presence of quaint Nantucketers, by contrast, keeps them available to be viewed and enjoyed by sophisticated summer tourists. Without ever invoking the word "quaint," *Nantucket Characters* crystallizes its visual style.

QUAINTNESS AND EXHAUSTION

Nantucket's quaint old men are both relics of the past and a warning to future generations: do not let yourself become exhausted. Given the visibility of decaying infrastructure and quaint old men on the island, Nantucket might have been a horror to the exhausted city worker—a place that on every corner offered a cautionary tale in the loss of energy. Paradoxically, a vacation on quaint Nantucket was seen by worn-out workers as an antidote to their own exhaustion. Vacationing became a way of restoring spent energy, and the language of saving labor and restoring energy through vacationing is an interesting mirror of the language used to describe the way that fossil fuels saved labor in the emerging culture of fossil fuel energy. The way that Northrup experiences vacation as a restoration of depleted bodily energy is evidence of the way that fossil fuel energy technology affected both the material technologies and infrastructure that changed the nature of work in fossil modernity, as well as the way that fossil fuels shaped the imagination and feeling of those workers. Or, as Bob Johnson puts it in describing the effect of massive fossil fuel consumption on American culture in the late nineteenth century, fossil fuel energy lay at the heart of "modernity's ecology": as a "flood of prehistoric carbon calories impressed itself in both conscious and unconscious ways on the modern body and mind, on our ways of being, knowing, and sensing in the world as Americans of different classes, races, genders, and conditions learned to embrace, absorb, and navigate the material manifestations and cultural potentialities of fossil fuels."[48] Much more attention is paid in scholarship about energy to the way that fossil fuels transformed work and infrastructure, but fossil modernity reshaped leisure, pleasure, and aesthetic appreciation at least as much. Vacations were promoted in the same way that new fossil fuel technologies were—as ways of saving labor and optimizing human performance for work. A person's capacity for labor—even the mental labor of desk work—was a type of account book that needed to be periodically balanced. As labor in US cities exhausted workers' minds, vacations were needed to balance energy expenditure and bring workers back into the black. Historian Cindy Aron argues that US vacation culture is defined by an anxiety about leisure in a culture that values hard work and diligence; the language of balancing the energy account book, of restoring enervated minds to "renew [the] old contests with the world," provided a moral justification for vacationing: a justification that became more and more necessary as fossil-fueled technologies made labor less physically exhausting for so many workers. The energetic metaphors for work and rest are enduring: Who has not felt their own "battery" running low or pleaded a few days off in order to "recharge"? These feelings and their expression through particular forms of rest and vacation have their roots in the energy transitions of the nineteenth century.

Nineteenth-century scientists had for a while observed that Americans—and, in particular, bourgeois northeasterners—were suffering from a particular form of nervous exhaustion and anxiety caused by the technological features of modern life. Neurologist George M. Beard consolidated these theories and gave the affliction the new name of "neurasthenia."⁴⁹ In 1881, Beard published *American Nervousness*, which expostulated on the causes of neurasthenia: "The chief and primary cause of this development and very rapid increase of nervousness is *modern civilization*, which is distinguished from the ancient by these five characteristics: steam-power, the periodical press, the telegraph, the sciences, and the mental activity of women."⁵⁰ Although Beard coined "neurasthenia," his analysis of nervousness was otherwise not novel in its description of symptoms or in its attribution of the cause of nervousness to new (and, notably, fossil-fueled) technologies such as the railroad, steam engine, and telegraph.⁵¹ Beard's analysis is notable, too, in the extent to which he developed the pathologies of nervousness and neurasthenia within the ruling idea of energy.⁵² For decades, scientists who had studied the nerves, from professionalizing neurologists to humbug mesmerists, had understood the nervous system through metaphors and theories of electricity, and Beard's analysis reflected the changes to electrical power production and industrial processes wrought by fossil fuels.⁵³ According to Beard, fossil-fueled energy regimes and steam-powered technologies such as machine manufacture, railroad travel, and steam-powered printing were both direct causes of nervousness and metaphors for the condition. For example, in one section, Beard characterizes nervousness as "nervelessness—a lack of nerve-force" and compares the human body to a battery that requires an adequate supply of stored energy.⁵⁴ Elsewhere, Beard elaborates his theory by analogizing the human nervous system to a steam engine that produces electricity to power a number of lamps, which are a man's organs and functional capacities.

> The nervous system of man is the centre of the nerve-force supplying all of the organs of the body. Like the steam engine, its force is limited . . . and when new functions are interposed in the circuit, as modern civilization is constantly requiring us to do, there comes a period . . . when the amount of force is insufficient to keep all the lamps actively burning; those that are weakest go out entirely, or, as more frequently happens, burn faint and feebly—they do not expire, but give an insufficient and unstable light—this is the philosophy of modern nervousness.⁵⁵

For Beard, fossil-fueled energy regimes created both the direct causes of nervousness and the language with which the new modern condition could be described. And while Beard's ideas about nervous exhaustion and modernity might not have been entirely novel, his articulation of "American nervousness" provides useful context for the appeal of Nantucket vacations for exhausted bourgeois northeasterners.

Not unlike George Beard, the twenty-first-century critic Franco Berardi explores the role of exhaustion in modern energy culture, and he even gives the alternate name "energolatria—the worship of energy" to modernity and cites the "imperative of growth" as "the defining characteristic of modern bourgeois economy.... The lack of growth, by contrast, is associated with exhaustion and all of its negative connotations."[56] "Exhaustion" in energy culture and criticism generally, and as Berardi uses the term here, has to do with the large-scale societal and environmental problems of coal and oil and the corresponding decline in economic growth that would likely attend resource depletion. But energolatria characterizes physical and mental discipline in the individual human body, too, and fossil modernity changed the way people understood their body's needs. Vacation cultures rose up to counter the specter of exhaustion in the body of the American worker. Here, too, as in accounts of the work of old energy regimes and infrastructures in the story of fossil modernity, accounts of the physical exhaustion and nervousness caused by modernity focus exclusively on ascendant technologies such as coal and steam. As with other modern experiences, nervousness exists in relationship to obsolescing energy regimes.

Traveling to seaside resorts and other watering places for rest and restorative cures was an activity with a long heritage, even in the nineteenth century. In his treatise on European cultural perceptions of the sea, historian Alain Corbin dates the beginning of European seaside rest cures to the middle of the eighteenth century and writes that sea bathing, in particular, engendered health and a particular form of self-knowledge: "At the edge of the ocean, modern man comes to discover himself and to experience his own limits in the face of the ocean's emptiness and the availability of the shores."[57] Corbin's account offers a useful genealogy for the form of vacationing that took hold in Nantucket and along the northeastern shore. A Nantucket vacation was the latest, locally and culturally specific eruption of a long tradition of traveling to the seaside for rest and health: one marked by the emerging energy logics of the fossil fuel transition. The development of seaside tourism in New England watering places such as Nantucket in the mid-nineteenth century coincided with the emergence of the word "vacation" in US print culture to describe a particular type of restorative trip.[58] Historian Cindy Aron cites an emblematic prescription for a seaside summer "vacation," named as such, in a *New York Times* article about the New Jersey seaside in 1855.[59]

Beard's account of American nervousness also helps us understand why places such as Nantucket that served as sanctuaries for sufferers of nervousness did not become shameful or secretive sites, the way other centers for medical treatment, such as asylums, would. This is because nervousness is, for Beard, the epitome of whiteness. The premise of Beard's pathology is that nervousness is an affliction, and its foundational definition in Beard's account as "a lack of nerve-force" carries a negative valence, but the total effect of Beard's account is of admiration and pride for its victims. Beard repeats throughout the account that the chief cause of American nervousness is "modern civilization," but if it is true that the conditions of modernity and civilization make one nervous, then it is also true for Beard that being nervous is proof of one's modernity and civilization. The physiognomy associated with nervousness is what Beard calls a "fine organization"—as compared with the "coarse"—and is marked, among other features unmistakably racially coded white, by "fine hair."[60] And the "fine organization" that signals a predilection to nervousness is, in turn, "the organization of the civilized, refined, and educated, rather than of the barbarous and low-born and untrained."[61] Beard reinforces the racial hierarchy of nervousness in every way, in one spot even according nervousness a quasi-monetary value: "Had we no other barometer than this, we should know that civilization was paid for by nervousness, and that our cities are builded [*sic*] out of the life-force of their populations."[62] Nantucket's reputation as a resort uniquely suited for nervous Americans is part of the source of its cultural prestige and its enduring whiteness.

Accounts of Nantucket vacations in the nineteenth century register the effect of the place on the bodies of exhausted vacationers. Several accounts describe Nantucket as an excellent place to sleep on account of its quiet and the unquantifiable soporific effects of sea air. Of all the places on Nantucket to achieve restfulness and better health, Siasconset (sometimes shortened to 'Sconset or Sconset), the quaintest village on the island, is the best place of all. "But perhaps the quaintest place on the Atlantic coast is the little village of Siasconset on the southeast corner of the island. . . . The village is a paradise for children, and a haven of rest for invalids, wearied brain-workers, and tired-out business men."[63] A. Judd Northrup, who wrote the narrative *'Sconset Cottage Life: A Summer on Nantucket Island* (1881) and whom we encountered earlier as a bored museum visitor, recounts his restful stay in Sconset in visceral detail. Upon his arrival on the island, he found that all he wanted to do for the first few days was eat and sleep: twin phenomena brought about by that mysterious sea air. But he quickly forgives himself and then credits the place for these awakened appetites: "I thought even this was worth something to the worn-out men who came hither for rest and recuperation—to sleep, to eat, and then to eat and sleep again. The nervous, headache-y, dyspeptic toiler at office desk, the head-weary of every calling, need these humble good things in their lives quite as vitally as a rejuvenation of their sentiments, a waking up of their enthusiasms or a kindling anew of youthful poetic fires,—but then at 'Sconset they may have all these and the slumber, too."[64] The type of worker who benefits from a Nantucket vacation, according to Northrup, is the office worker, the "toiler at office desk," who is not physically exhausted like a manual laborer but is exhausted nonetheless. For Northrup, the maladies of the worn-out office worker are physical even though the office worker's labor is not particularly strenuous: exhaustion manifests in headache and dyspepsia. And the simple cure that Nantucket offers, of eating and sleeping in sea air that stokes the appetite for both, restores energy to the exhausted worker. Northrup describes the process of recovering from the maladies of office work through a Nantucket vacation in explicitly energetic terms: "waking up" and, especially, "kindling anew of youthful poetic fires." Vacation enlivens the body of an exhausted office worker in the way that mirrors the combustion required for energy production.

According to the emerging tourist literature, Nantucket was uniquely suited to cure the ills of nervous and exhausted city workers *because* of the rosy hue of quaintness and retirement that visitors attached to the sight of its declining industry, depopulated town, and aged infrastructure and inhabitants. A depopulated postindustrial town is the perfect place to seek health and restoration, suggests the guide *The Old Colony; or, Pilgrim Land, Past and Present*, published by the Fall River Line and Old Colony Railroad in 1887:

> Indeed, the dreamy, quiet, conservative conditions which now obtain in every part, have perhaps contributed more towards making this place the famous watering-place it has become, than her citizens would be willing to allow. The absence of the rush and whirl of competitive business operations, the free devotion of every natural feature and facility and advantage of the island to the interests of the summer sojourner and pleasure-seeker, the union of relations between the residents and the tourist and visitor in these directions, have done more in developing those conditions which in conjunction result in the finest and most desirable summering establishments, than could possibly have been wrought otherwise.[65]

Quaintness was an antidote to nervousness, for reasons that Beard's theory of nervousness elucidates. Among the conditions of modernity that cause nervousness, according to Beard, are the prevalence of clocks, which make people newly conscious of the time, and the proliferation of telegraph technology, which speeds up commerce and the fluctuation of value. Vacation life on Nantucket, by contrast, is marked by the "absence of the rush and whirl of competitive business operations." The distance between regimented regimes of clocks and telegraphs and the town crier's idiosyncratic news reports, encountered serendipitously, help explain the prominence of the town crier in tourist accounts of the island.

In the late nineteenth century, as now, the rich visitors and comparatively poor permanent residents of Nantucket led starkly unequal lives: an inequality that visitors hastily but imperfectly work to elide by fabricating myths of common experience. The Old Colony guidebook speaks euphemistically of "the union of relations between the residents and the tourist and visitor in these directions," as if the decline of whaling and the depopulation of the island were situations that the impoverished islanders had chosen in order to pursue a restful retirement. But the island's businesspeople had specifically and materially promoted tourism and made the whaling port over into a "watering place" in order to find a new source of profit; this, rather than the fortunate alignment of interests among resting souls, accounts more for the "union of relations" between residents and vacationers.[66] It required a willful and profound misreading of the local island economy for visitors and promoters to assume that aged Nantucketers enjoyed precisely the same benefits of health and pleasure from living on the island as did visiting vacationers. And yet, many other accounts romanticize the presence of underemployed old-timers as further evidence of the healthful benefits of life on Nantucket. "A gentleman visiting the island was surprised at finding so many aged people there, and remarked to an old sea-captain, that he 'wondered if anybody ever died on the island.' The captain answered 'Die? Never! They merely dry up and blow away!' The town is very healthy indeed, there being at the present time several persons over ninety years of age."[67] The tourist guides project the vacation values of health and restored labor onto the aged island residents, misreading the prominence of old people and absence of younger ones as a sign of the island's health rather than of its depopulation and the lack of employment. Such misreadings are necessary in order to find pleasure in quaintness, as in the misreading that finds empty warehouses quaint rather than signs of economic evacuation. Enjoying the pleasures of quaintness meant repressing the knowledge of death of all kinds (the death of Nantucketers, the death of industries) and sublimating that knowledge into quaintness.

CODA: ALLAN SEKULA AND THE QUAINT SEA

In the same decades when tourists began flocking to Nantucket, something of the quaintness that attached to small seaside places in the late nineteenth century drifted out from shore and spread to the sea itself. In the transition to fossil fuels, whaling was not the only maritime industry that came to be known, at least among a particular set of US elites, as obsolete. The sea itself came to seem obsolete and quaint, no longer a place of work or essential functioning. Or so goes a commonplace and enduring narrative about maritime history. To this point, I have discussed the way that Nantucket quaintness reveals the paradoxical anxiety about and attraction to exhaustion as one of the emerging logics of energy in the transition to fossil fuels in the late nineteenth century. The case of Nantucket quaintness also illuminates a change in the histories and perceptions of the sea: the emergence of the quaint sea.

American historians in a long lineage ranging from Frederick Jackson Turner to today hold that in the last half of the nineteenth century to the present, the sea became less important as a space of economic power and cultural production than the inland "frontier," where fossil capital moved along with settler colonialism over the expanding territory of the western continent. In 1893, while Nantucket tourists hunted antique shops for harpoons, Frederick Jackson Turner articulated the long-enduring "frontier thesis" of US history, holding that (white, masculinist, settler) US identity was forged on the western frontier. It seemed to many people ashore, and to historians looking back on the period, that the sea became "invisible" or "forgotten." Artist and critic Allan Sekula charged mass culture with a condition he called "forgetting the sea," a diagnosis that literary critic Margaret Cohen has elaborated upon and further pathologized as "hydrophasia."[68] Most of these historians and critics have focused on the sea as a space of work, but even Alain Corbin, who studied seaside leisure and the way that European artists and writers represented the sea, ended his survey in 1840.[69] It has been the work of the field of oceanic studies since that time to draw the attention of distracted readers back to the sea.

The "forgetting of the sea" after the mid-nineteenth-century can be attributed not to some wider popular amnesia but to a particular historiographical tendency to focus on the age of sail for sentimental or aesthetic reasons. Maritime historians for a long time cleaved to the notion of a golden age of American seafaring in the first half of the nineteenth century, during which time the United States outstripped the previous Western maritime power, Great Britain, in mercantile shipping, naval power, and whaling and rose to an otherwise often unspecified form of maritime dominance.[70] In his *Maritime History of Massachusetts* (1921), considered a classic work of maritime history, Samuel Eliot Morison announces his refusal to investigate maritime history beyond the age of sail: "I have chosen to catch the story at half flood . . . and to stay with it only so long as wind and sail would serve. For to one who has sailed a clipper ship, even in fancy, all later modes of ocean carriage must seem decadent."[71] For Morison, maritime history beyond the age of sail held an aesthetic, and even a moral, repugnance—a bias that continues to haunt US maritime history and heritage, which often acts as if the history of the sea itself ended with the age of sail. (In chapter 5, I explore the nostalgic wood-and-sail bias of maritime history and heritage as one of the enduring legacies of *Moby-Dick*, the Melville Revival, and whaling history in design and visual culture.)

But there are other, more legitimate, stories to tell about how popular cultural representations of the sea shifted along with technological changes in shipping and, I would argue, new configurations of energy and infrastructure in the fossil fuel age. One of the most compelling is the explanation Allan Sekula offers in *Fish Story* (1989–95), a long-term project in the multimodal form he calls "critical realism." *Fish Story* took its form in a series of museum exhibitions and in a book that is much more than a catalog; it is a long-form and deeply researched essay interspersed with Sekula's photographs. Through *Fish Story*, Sekula argues that the sea is a space that fellow "intellectuals" have forgotten in their peripatetic airplane travels, their rush to navigate "cyberspace," and their experiences of instantaneous long-distance communication through digital technology. Their forgetfulness is a grave mistake, according to Sekula, because the sea is the best space from which to understand the logics and flows of global capitalism, the material movement of goods and people around the globe, and the hidden forms of exploitation and violence to laborers and migrants that forgetting enables. To counter the forgetting of the sea, Sekula documents maritime space in the late twentieth century with the photographs and essays about container ships collected in *Fish Story* and in a constellation of related films and other projects.

Like other scholars of the sea, Sekula offered an argument for why the sea became invisible. Unlike US historians who imagine that the colonial expansion in the western part of the continent turned national collective attention away from the sea, or unlike those who turn in distaste from maritime history after the age of sail, Sekula dated the "forgetting of the sea" to the 1950s with the development of containerization: the process of transporting commodities and manufactured goods in standardized "containers" that could that could be moved seamlessly—that is, by mechanized cranes and robots and with a minimum of human labor—from factory to ship to truck to consumer. According to Sekula, containerization reshaped port cities and radically changed the way that people outside the shipping business interact with maritime trade: goods that had once been available to see, hear, and smell are cordoned off in containers, while commercial container ports have moved out of metropolitan centers to the edges of cities. Through containerization, landbound folks and even harbor residents have become alienated from the material, lived reality of maritime trade and find themselves vulnerable to imagining away the lived experience of maritime trade into the abstractions of logistics and markets. Sekula counters the forgetting of the sea with vivid images of ports and container ships and with compassionate portraits of the people who live and work at sea.

Sekula writes with contempt for the people he calls "intellectuals" or "late-modernist elites" who cruelly ignore the privations of maritime labor or enjoy the sea merely as a "bourgeois reverie on the mercantilist past."[72] In so doing, Sekula addresses an audience that in many ways resembles the subjects of this chapter: the wealthy, white, urban quaintness-hunting tourists who flocked to Nantucket in the late nineteenth century. Sekula's charge can easily be refitted as an alternate definition for quaintness as I explore it here: a bourgeois reverie on past energy regimes. Sekula continued the work of *Fish Story* in an "essay film" he coproduced with filmmaker Noël Burch called *The Forgotten Space* (2010) that updated the insights of *Fish Story* for a world in an economic recession. In an accompanying essay, Sekula and Burch enumerate the two "myths" that enable people to forget the fundamental importance that oceanic trade holds for global capitalism: "The first myth is that the sea is nothing more than a residual mercantilist space, a reservoir of cultural and economic anachronisms. The second myth is that we live in a post-industrial society, that cybernetic systems and the service economy have radically marginalized the 'old economy' of heavy material fabrication and processing. Thus the fiction of obsolescence mobilizes vast reserves of sentimental longing for things which are not really dead."[73] More than 100 years after the first tourist boom on Nantucket, Sekula and Burch describe the "forgetting of the sea" in the language of quaintness—"anachronism" and "obsolescence"—and they charge those who forget the sea or celebrate its quaintness with a delusional "sentimental longing for things which are not really dead." Sekula's implicit criticism of the maritime quaintness aesthetic presages critiques of Detroit ruin porn: both are discourses that ignore the lived realities of people who work in and inhabit places relegated by the touristic gaze to the obsolete past.

But Sekula's stance on maritime quaintness is ambivalent. Despite Sekula condemning the forgetting of the sea, and the quaintness discourse that is one of its processes, he indulges in nostalgia for the harbor spaces of his youth. Sekula even calls his own childhood perception of harbor spaces "quaint": "Growing up in a harbor predisposes one to retain quaint ideas about matter and thought. I'm speaking only for myself here, although I suspect that a certain stubborn and pessimistic insistence on the primacy of material forces is part of a common culture of harbor residents."[74] A few lines later, Sekula reconstructs the rich sensory environment of the harbor spaces he remembers by way of contrast with the sense-deadening container port that abets the abstractions of capitalist imagination: "Goods that once reeked—guano, gypsum, steamed tuna, hemp, molasses—now flow or are boxed. The boxes, viewed in vertical elevation, have the proportions of slightly elongated banknotes."[75] The catalog of sensorily rich, "reeking" substances—"guano, gypsum, steamed tuna, hemp, molasses"— erupts unbidden from Sekula's description of bygone harbor spaces, like the lost world that magically unfolds for Marcel Proust after that first taste of the madeleine. For Sekula, considering the sea and harbor spaces as quaint is the delusional attitude of people who can afford to avoid its material reality and who lose the opportunity to understand the sea's centrality to global capitalism. But Sekula's "quaint" belief in "the primacy of material forces," forged in the harbor spaces of his youth, is arguably the most concise definition of his materialist ideology and practice. Attending to the material, lived reality of oceanic transport, labor, and migration is politically urgent. But to make his point, Sekula indulges in reminiscences that sound remarkably like nineteenth-century accounts of Nantucket quaintness. As I discussed above, Nantucket tourism literature often listed and lamented the bygone sights and sounds of a busy port, recalling "the cheery din of the cooper's hammers and the ring of the blacksmith's anvils resounded on all sides" and mourning the quieter present, when "the tools of the mechanic are silent, and the bustle of traffic no longer crowds the streets"[76] A touch of Nantucket quaintness haunts *Fish Story* in other respects, too: the critique alludes to the specific history of nineteenth-century commercial whaling with luminous photographs of a museum's miniature whaleboat model and a scrimshawed sperm whale tooth featuring an erotic scene.[77] The photographs are interspersed in a section of *Fish Story* that otherwise offers intimate portraits of workers and the shipboard and port spaces they inhabit. In this context, the artifacts of whaling history serve to engender a viewer's empathy toward maritime laborers past and present. The interruption of contemporary photographs with images that in most other contexts would read "quaint" reveals the uses of quaintness, even for Sekula. Quaintness is a mode of understanding the sea and its history that not even Sekula can entirely resist.

Sekula's ambivalence about quaintness might be related to his own crimes of forgetting and, in particular, to his unwillingness to reckon with race. Christina Sharpe notes that Sekula invokes the Middle Passage in *Fish Story* without ever referring to "the planned disaster that is known by that name."[78] And although a Black woman is one of the subjects of *The Forgotten Space*, "the film operates within a logic that cannot apprehend her suffering."[79] Sekula seems to understand the sea as a space of leisure as well as of aesthetic experience from the singular point of view of whiteness, although he does not acknowledge that he may be generalizing a universal experience from that specific racial experience. Quaintness on Nantucket was and is connected to the discourse of whiteness, because quaintness draws on racialized understandings of labor and exhaustion. But Nantucket and other New England resorts never were places without Black people, Indigenous people, and other people of color—and quaintness is not the only way of being in these places. In research about Black communities on Nantucket, Tiya Miles identifies orientations to place and community that are different from quaintness; her piece is aptly titled "Nantucket Doesn't Belong to the Preppies." She sees in the Black communities of Nantucket a powerful history of collective caretaking and a model for climate resilience in the present and future.[80]

This chapter continues the work of such scholars as Hester Blum, Alain Corbin, Allan Sekula, and Christina Sharpe to document the changing material and infrastructural realities of the sea, the coastline, and harbors; to witness the lived experience of those who live, work, rest, and recreate at sea and alongshore; and to account for the cultural work of the sea in widespread popular culture. My intention here, and in the book in general, is to provide a counterargument to the "invisibility" thesis of oceanic culture—or, at the very least, a prehistory to the "forgetting" that Sekula dates to the mid-twentieth century. Rather than make a claim about the "disappearance" or "forgetting" of the sea at a moment of technological or infrastructural change, I have shown that the sea was more visible than ever, even to those who did not labor in ships or ports. At least some nineteenth-century observers who beheld the sea—particularly from quaint places such as Nantucket—beheld a sea that began to appear like something from the past. They did not turn their backs on the sea; they called it quaint.

Major technological changes, such as the intensification of fossil fuel consumption in the nineteenth century or containerization in the twentieth century, change the scripts that govern the movement of people in and through space. Changes in energy and technology necessarily change the way environments and infrastructural systems are widely represented and understood. In focusing on late nineteenth-century Nantucket, we can see how a specific class of tourists negotiated the emerging logics of energy and exhaustion in fossil modernity and understand that the quaintness they so prized on Nantucket revealed changing perceptions of work, leisure, and even the sea, too. Nantucket is not alone in the culture of quaintness that emerged in its transition to tourism, but I hope to resist the impulse to generalize too broadly based on the experience of a limited group of privileged East Coast urbanites. Rather, I hope to provincialize quaintness and to show that it is a particular style of thinking about energy and the sea that emerged at the intersection of declining whaling infrastructure and emerging fossil fuel infrastructure.

The mass movement of tourists to the quaint seaside was only one possibility engendered by fossil modernity. The next chapter explores how the intersection of old, decaying whaling infrastructures and fossil-fueled infrastructures changed the landscape of the great US inland and made the ocean itself mobile. The sea and even its creatures would become spectacularly visible, even from the distance of a thousand miles.

3

Pioneer Inland Whaling
A Whale on a Train, a Ship Called *Progress*, and the Transformation of Whaling Culture in the Inland United States

The same coal-fired steamships and railroads that brought tourists to the faded whaling port of Nantucket also carried whaling culture in the opposite direction: from the oceans to the lakes, prairies, and settler cities of the mid-continent. The cultural afterlife of US commercial whaling in the last decades of the nineteenth century moved to the unlikely site of Chicago, which was then, according to landscape historian William Cronon, "the greatest metropolis in the continent's interior."[1] Chicago had grown at an astonishingly rapid pace in the mid-nineteenth century, quickly changing its identity from a midcontinental entrepôt to the center of US commerce and production. This chapter explores the inland instantiations of whaling culture through the story of two spectacular exhibitions that traveled to Chicago. First, from 1880 to 1882, two impresarios toured the decaying body of a dead whale, called the Prince of Whales, throughout the United States as a kind of circus exhibition. After its death in the North Atlantic, the whale traveled by rail from Boston Harbor to Chicago for its grand debut. Its unlikely tour continued along rail lines throughout the Midwest and South to curious audiences, most of whom had never seen the sea. One decade later, a group of industrialists from Chicago bought a wooden New Bedford whaler called the *Progress* and brought it through the inland canal network to Chicago for the World's Columbian Exposition in 1893.

The unlikely inland voyages of the Prince of Whales and the *Progress* occurred at a moment of chiastic crossing when fossil fuel consumption was on the rise and the whaling industry was on its steep decline—in the very middle of what Christopher Jones has called the "shift from an organic to a mineral energy regime."[2] By the winter of 1881, New England ports were still launching whaling voyages, but the number of new voyages was decreasing rapidly each year.[3] In the 1840s, the United States launched 2,363 whaling voyages, but during the 1880s, when the Prince of Whales toured the country, only 736 voyages departed American ports.[4] And in the 1890s, only 426 voyages left US ports.[5] The whaling industry was implicated in the extraction and consumption of coal, too. Jones chronicles the "positive feedback loops" of the mineral energy regime, noting that by the mid-nineteenth century, the tools used to extract coal, along with the ships and trains used to deliver coal, all ran on steam engines that consumed coal.[6] The US whaling fleet was by and large a fleet of wooden sailing ships. As the mineral energy regime and the use of steam technology intensified, wooden sailing ships became obsolete right along with whale oil.

It seems unlikely that the impresarios of the whale show could ever have commissioned a whaling captain to bring home a whale's body during the boom days of the whale oil economy; only because whaling had entered its obsolescence was the Prince of Whales available for a sideshow tour. Had the price of whale oil or other whale commodities been higher, the owners of the whaling ship *Progress* would never have countenanced the vessel's exit from the whaling grounds and its transformation into a museum.

The historical developments that often focus historians' attention on Chicago and the American West in the late nineteenth century are often cited as the same forces that ended the heyday of US maritime culture: the settlement of the western frontier through the genocide and forced removal of Indigenous people, the transformation of the landscape into a site of productive agriculture, the territorial expansion and incorporation of western lands into the US nation, and increasing fossil fuel extraction that caused whaling to decline and fueled the expanding overland railroad system. Hester Blum, for example, names 1850 as the end of the "so-called golden age of American shipping (1815–1850), after which the commercial and popular imaginations of the nation were increasingly directed along overland western routes."[7] Blum's argument for the tailing off or transformation of a certain type of maritime narrative is authoritative, although I would argue that the movement of commerce and culture "along overland western routes" *included* maritime culture and oceanic objects. The new infrastructures of overland western travel, commerce, and culture enabled people who had never seen the ocean to have direct sensory encounters with the ocean.

Oceanic culture by no means disappeared as that so-called golden age of maritime history ended. Maritime culture, and whaling culture in particular, *moved*. When whaling history moved, it traced new histories and forecasts of energy. The transplanted whale and whaling ship served in the American West of the late nineteenth century as yardsticks for measuring fossil modernity's progress: the comparative largeness, heavy weight, and complexity of the logistical ventures that fossil modernity makes possible. In order to see and account for maritime culture as it moves, new and more capacious definitions of "infrastructure," "media," and "culture" are called for. To that end, I analyze the Prince of Whales sideshow and the whaling ship *Progress* as media: *whale media* and *ship media*.

In this chapter, I draw on media theory in order to analyze these objects, inquiring into the messages they transmitted, the means of their transmission, and the remediations that whale and ship underwent. The approach I take to describing these objects is informed, too, by practices of material culture. Media theory augments my study, though, since neither whale nor ship is still extant for direct observation. Theories of remediation, proposed by J. David Bolter and Richard Grusin and taken up by media scholars since, afford ways of thinking about the Prince of Whales and the *Progress* as objects that change under the pressure of the economies, infrastructures, and energies changing around them.[8] The whale and the ship expand the theory of remediation, which does not apply only to the new media that Bolter and Grusin put at the center of their study but which also helps describe the way old infrastructures change and flicker in and out of visibility.

Media theory also holds useful affordances for thinking of media across scale, from the finest details of design to the largest pieces of continental infrastructure. Nicole Starosielski and Lisa Parks have advocated for media studies to "[adopt] an *infrastructural disposition*": an approach that necessarily brings media studies into collision with energy and oceanic humanities, given the energy resources and the vast infrastructures—often transoceanic or submarine—required to distribute both analog and digital media.[9] My study of whale media and ship media is one, perhaps, of infrastructure with a media studies dimension. A media studies approach to the whaling ship *Progress* is especially fitting, because a ship was the subject of history's first work of remediation theory. The ancient Greek philosopher Plutarch confronted the "ship of Theseus" paradox, asking whether a ship whose timbers have all been replaced over time is still the same ship it was when it was built.[10] As with all remediated objects, the answer is, of course, yes and no. The extravagant and even absurd experiment of removing oceanic subjects from their environment is a real-life, historical example of "conceptual displacement," which Melody Jue defines as a speculative methodology for thinking about media through the milieu of the ocean. Conceptual displacement means imagining media under the ocean. The speculators who brought a whale and a ship did something equally speculative: they brought oceanic beings to land. Jue writes that "conceptual displacement [is] a method of defamiliarization to make our terrestrial orientations visible."[11] The exhibitions of the Prince of Whales and the whaling ship *Progress* were exercises in conceptual displacement, and their stories do, in fact, reveal a history of our terrestrial orientations. Their stories also show the surprisingly intimate connection between the terrestrial and oceanic, and they reveal a wide US territory positively swimming with oceanic artifacts.

Some features of energy culture remain constant across different energy resources and extractive regimes: energy is entangled with extraction, infrastructure development, and settler-colonial violence. But the new forms of energy added to the US landscape in the late nineteenth century radically reshaped the North American continent and collapsed the space separating the US inland from the sea. The oceanic infrastructures of whaling, whales, and ships stood as powerful, rotting metonyms for the organic energy regime in eclipse. Chicago, by contrast, was known for its railroad hub and the mighty meatpacking industry that condensed the energy of the western continent into animal protein. Chicago's terrestrial infrastructures stood for fossil modernity in ascendence.

THE PIONEER INLAND WHALING ASSOCIATION: A STORY

In late November 1880, George H. Newton, a sometime real estate agent and lawyer from Monson, Massachusetts, received the telegram he had been awaiting: a whaling ship bound home to Provincetown had caught him a whale. Newton raced to Provincetown to claim the enormous body. He arranged for a towboat to float the whale's body to Boston and for a drydock in Boston Harbor—the kind used to raise ships out of the water for repair—to lift the whale's body out of the water. On land, the whale's body was sliced open, and its entrails were lifted out and replaced with a combination of ice and salt.[12] Within a few days, a jury-rigged crane system moved the dead whale from the dock to two train cars (the whale was too long to be supported on one) and prepared for shipping.

For almost two years after the whale's body came out of the cold water in Massachusetts, the whale—or something like it, anyway—toured by train through the United States as a traveling exhibition: an exotic spectacle from the sea that inland customers would pay to see. Newton had wanted to partner with P. T. Barnum, but the famous exhibitor had declined; one of Barnum's agents replied that the great showman's "time is so taken up that he could not give such a speculation any attention."[13] Newton partnered instead with Fred J. Engelhardt, a Chicago-based sports promoter and journalist. Newton and Engelhardt called themselves the Pioneer Inland Whaling Association, and in late December 1880, they shipped the whale west to Chicago.

The whale went on view at Chicago's Exposition building in January 1881. Throughout January 1881, the show drew huge crowds. By Newton's count, some two thousand to four thousand visitors came down to see the whale every day for the first few weeks of the whale's stay in Chicago. Part of what made the exhibition of the whale possible during those first few months in the Midwest was an unseasonably cold winter. Newton and Engelhardt were lucky with the weather: by all accounts, it was a cold January, and the whale remained frozen.[14] A few weeks later, the same paper noted the luck of the cold weather: "The only manager who has particularly rejoiced at his cool reception in Chicago this season has been the proprietor of the 'forty-ton whale.'"[15]

After Chicago, the whale show traveled on to Milwaukee, St. Louis, Cincinnati, Louisville, Pittsburgh, Philadelphia, Cleveland, Toledo, and Detroit. By April in Philadelphia, the whale began to thaw and Newton worried about "the bird," his nickname for the whale: "If the bird gives out we go out of business for a while."[16] Suffice it to say that the bird gave out, and in the spring of 1881, Newton and Engelhardt embarked on series of increasingly desperate attempts to preserve the whale and keep it on the road. During the whale show's stay in Cleveland later in April, the proprietors hired a team of butchers to remove the remaining flesh from the inside of the whale. Then they built a wooden box around the whale in order to coat its body with a chemical substance. The procedure was not concealed from the public but rather publicized and romanticized: "Captain Newton, one of the chief spirits in the Pioneer Whaling Association, became possessed with the idea that if people in former ages could preserve bodies, there was no reason why Yankee ingenuity could not discover this system, or at least invent one just as good; so he, in conjunction with Boston parties, commenced a series of experiments directly after he became the owner of the great sea monster which is now generally known over this country as the 'Prince of Whales.'"[17] It appears that the chemical treatment failed. Within less than two months of the whale's treatment in Cleveland, the whale show was thrown out of Detroit on account of its excessive smell. In the summer of 1881, Newton traveled home to Massachusetts, and Engelhardt set out with the whale to a former logging camp near Bay City, Michigan, for the summer. Engelhardt hired Detroit taxidermists and a team of carpenters to work on reconstructing the whale. Engelhardt called the site Camp Baleine and peppered the newspapers with stories of the whale's marvelous preservation. Although the retreat to northern Michigan reflected the failure of the first "embalming," Engelhardt publicized the whale's whereabouts and began planning a new stage of the tour.[18]

At that camp in northern Michigan, Engelhardt and his team remediated the whale: addressing its smell and, as the word "remediation" suggests, made the whale from one medium into another. The whale's body was straightened, blown up, evened out, stretched out, its surface made smooth; incisions and mutilations made by the harpoons and marauding visitors were removed and replaced by an even, fabricated surface. Engelhardt wrote to Newton: "We are here—in saying that I say almost all that can be Said.—Mosquitos, flies, bugs and snakes predominate and form the largest part of the atmosphere. . . . The work is terrible—you have no idea—."[19] After the whale's dreadful sojourn at Camp Baleine, Engelhardt and the whale set out again to Chicago for a return exhibition in September 1881.[20]

From there, the whale continued toured points west and south for over a year, but the Pioneer Inland Whaling Association began to unravel. In October 1881, Newton sold his share of the whale show back to Engelhardt for $10,000 but remained entangled with Engelhardt and other workers in the show in legal disputes and heated arguments over unpaid wages.[21] Through the winter of 1881 and spring of 1882, the whale rolled south through Georgia to New Orleans and then north through the Carolinas to Washington, DC, by May. At some point, Fred Engelhardt left the show, unenriched. Reports from the late 1880s place Engelhardt in California, where he was managing a team of racing lady bicyclists, having returned to the promotion and management of human bodies.[22] The whale show was taken over by Paul Boyton, a minor celebrity in aquatic sports, famous for his invention and use of a buoyant rubber "life-saving suit": the aquatic adventurer joined forces with the aquatic monster.[23]

By the summer of 1883, newspapers report that the whale had been returned to Chicago, abandoned on a sidetrack, and widely decried as a fraud. Accounts of the whale circulated in newspapers through the end of the decade, mainly to report what marauders had found when they started taking apart the abandoned whale exhibition: badly preserved whale flesh stitched to paper muslin, stretched over a large wooden frame and stuffed with barrels and moss. "That things are not always what they seem, is exemplified in the case of this whale. The skin and tail of this monarch of the vasty deep was all that it purported to be, but its frame, alas! was of iron and hickory, and its flesh of sawdust and other deceptive light weights."[24] Reports of the whale fraud bubbled up for years, claiming that the whale on the train, in fact, never was.

WHALE MEDIA

The story of the whale show offers a knot of epistemological and ontological problems. Media theory, if we can begin to see the whale itself as a medium, offers useful tools for interpreting the messages that the whale transmitted, the means of their transmission, and the changes that the whale-as-medium underwent during its tour. A dead whale show is an unconventional medium, to be sure, but not one without precedent in media studies. Media theorists since Marshall McLuhan have put pressure on the concept "medium," examining everything from automobiles, human bodies, railroads, moving walkways, stones, oil, ice, brainwaves, and insect swarms under the rubric of media.[25] Researchers in the environmental humanities have drawn on media theory in order to better understand the material basis for and circulation of natural substances in and around conventional cultural texts such as books, film, music, and digital media. Stephanie LeMenager has theorized "oil media" in order to describe the complete enmeshment of all cultural production within the petroleum economy, and she has written, too, about infrastructure as media.[26] In a similar move, Nicole Shukin, in *Animal Capital*, explores the enmeshment of modern cultural production with the mass slaughter of animals through a theoretical figure she calls "rendering," drawing out the double entendre to show the relationship between "the mimetic act of making a copy . . . *and* the industrial boiling down and recycling of animal remains."[27] Among the many sites of animal capital she explores is photographic film, which beginning in the 1870s was coated with animal gelatin: "In the translucent physiology of modern film stock—in its celluloid base and its see-through gelatin coating—it is possible to discern the 'two-layered' mimesis through with modern cinema simultaneously encrypts a sympathetic and a pathological relationship to animal life. Film thus marks a site where a contradictory logic of rendering is daringly, yet inconspicuously, flush."[28]

Oceanic studies draws productively from media studies, too: Melody Jue proposes the ocean as a site for better understanding the "role of milieu in the fields of media studies, literary studies, science and technology studies, and the environmental humanities"; Stacy Alaimo's "pelagic posthumanism" was incited by new visualizations of (living) jellyfish and other pelagic creatures; and Hester Blum's past and ongoing work demonstrates the coevolution of maritime labor and textual production.[29] The payoff of practicing an environmental or "green" media studies is that this approach reveals the inseparable entanglements of form, theory, and material: it unites the Marxist and Foucauldian practices of old materialists, who can interrogate the material structures that underpin discursive and cultural power, with the anti-anthropocentric ethics and imagination of new materialists, who unseat human hubris by locating nonhuman agencies and exploring the vicissitudes of materiality in its difference and variation.

The media-infrastructure approach is especially resonant in the case of the Prince of Whales, whose message was inseparable from its materiality and whose means of transmission was a continentwide assemblage of trains, railroads, and human and animal labor teams. And as the foregoing narrative demonstrates, the whale itself was constantly remediated: the whale was transformed through death and putrefaction and also through Newton's and Engelhardt's various attempts at arresting putrefaction through freezing, chemical treatment, and taxidermy. Jay David Bolter and Richard Grusin have theorized remediation as a way of understanding new digital media, but their ideas enable new understandings of the whale show, too.[30] Bolter and Grusin describe remediation as "the representation of one medium in another."[31] In the case of the whale show, the whale was part of the original medium that was remediated, in pieces and over time, into ice, salt, moss, chemicals, wood, steel, and new organic material that emerged through the body's putrefaction. Another integral component of the whale-show-as-medium was the complex transportation network that transmitted it across the country and the structures of display that were built anew at every stop along the way.

Remediation has a "double logic," according to Bolter and Grusin, demonstrating "our culture's contradictory imperatives for immediacy and hypermediacy": immediacy offering "the transparent presentation of the real" and hypermediacy invoking "the enjoyment of the opacity of media themselves."[32] The whale show demonstrated both imperatives fully: it allowed visitors a rare experience of immediacy through sensory contact with a far-fetched specimen. And at the same time, the whale show offered visitors a chance to revel in the whale's hypermediacy: the logistics of the whale's transport from the North Atlantic all the way to Chicago and points beyond was a topic of constant promotion. Remediation is arguably a characteristic of many or even all cultural forms (and Bolter and Grusin acknowledge a genealogy for digital remediation going back at least several hundred years). What made Newton's and Engelhardt's whale special is that it offered immediate sensory experience with a pelagic creature that had been otherwise inaccessible, a whale that lived outside human habitation and whose rendered whale oil had been circulated, if not invisibly, then "inconspicuously," to use Nicole Shukin's word, through ordinary consumer spaces distant from the coast. Its hypermediacy was no less wondrous—its transportation halfway across the continent and the almost occult, death-defying powers Newton and Engelhardt claimed in their efforts to halt its putrefaction.

THE IMMEDIACY OF THE WHALE: PUTREFACTION AND SMELL

The immediacy of the whale—its availability to a multivalent sensory inspection—was touted as one of its major attractions. Promotional materials invited visitors to encounter the whale, and especially to encounter its yawning mouth—"the place where Jonah went in!"—in language that invites a type of transcorporeal experience. Visitors to the whale approached the whale at its head and walked alongside the body on a raised platform as an announcer called attention to various anatomical features. A reporter from the *Daily Inter Ocean* described the visitor's experience: "A visitor begins his observation generally at the head of the fish, looks into his capacious mouth, feels of the long, bony hair that supply the place of teeth, hunts for the eyes, the snout, and then the ears: walks along the side of the 'creature,' catches hold of the huge 'fin,' punches the monster in the side, as if to ascertain if it is ribless, and finally brings up at the tail of the huge fellow, where the broad flukes are spread."[33] This account chronicles a tactile encounter with the whale: the reporter describes touching the whale's baleen, taking hold of its fin, and punching its side. Some accounts describe the hole in the whale's carcass where they suppose the harpoon entered its body. Others described seeing large craters on the whale's body where flesh had been cut away. So many visitors attempted to carve out souvenirs that stern police officers, described as "Pinkertonian" in one account, were stationed nearby to guard the corpse.[34] Like a soft clay tablet, the whale's body recorded those moments of tactile encounter and the ongoing injury to its body.

On one hand, this encounter seems to instantiate exactly what Stacy Alaimo and Jane Bennett had imagined of encounters with dead fish: a mode of environmental perception that licenses human violence. On the other hand, the encounter demonstrates a moment of acute sensory awareness, constructing the dead whale as a vibrant if not a living body. Recent scholarship in the history of the senses has described senses such as touch and smell, especially, as embodied ways of knowing that establish boundaries between the body and the world and, at the same time, construct the body's porosity and bring it into contact with other bodies and the surrounding environment.[35] Offering a historiography of touch, Elizabeth

D. Harvey notices the persistence of Plato's theory of the flesh as "a creative interface between body and world," rather than as an unbreachable boundary between them.[36] That understanding of touch is borne out in the *Daily Inter Ocean* reporter's account of his encounter with the whale. The reporter recounting this visit describes an inventory of his own experience in parallel to an inventory of the whale's own sensory organs (eyes, snout, ears), as if making an implicit comparison between his own body and the whale's. The reporter's touch—albeit a punch—is offered up as the means of accessing and inquiring into the whale's interior.

Smell destabilizes hierarchies, too, revealing the vibrancy and agency of matter even more forcefully than touch can. Mark S. R. Jenner has traced a history of smell, noting that in many times and places, odors were understood to be material substances that entered human bodies and transformed them.[37] Historian Martin Jay describes the effect of Jenner's scholarship and the contributions of other historians of smell in strikingly oceanic terms: "[The history of smell] helps situate us in a more fluid, immersive context, where such stark oppositions [between subject and object] are understood as themselves contingent rather than necessary."[38] Where we cannot "follow the submersible," as Alaimo advocates, we can attend to sensory experiences can help us find that we are always already submerged: porous bodies immersed in a field of swirling material agencies.

The threat of encountering a putrefying body charged visitors' encounters with the whale with anxiety. Smell rather than sight indexed the whale's decomposition, and Newton and Engelhardt worked futilely to keep the smell at bay in order to keep the whale show profitable. An advance notice of the whale's arrival in Chicago before the exhibition's opening offers the promoters' reassurances: "[Engelhardt] also claims that beyond the natural smell of old ocean the big fish is odorless, and has not what Shakespeare calls an 'ancient and a fish-like smell.'"[39] And, at least in Chicago, the news reports affirmed the whale's relatively inoffensive odor. A *Daily Inter Ocean* reporter who visited the whale before the exhibition officially opened noted the smell, remarking that he "was met by a decidedly aggressive odor of fish and oil."[40] But others who commented on the whale's smell used mild terms: "The flesh, in color, resembles hog, and smells like seaweed."[41] Meanwhile, in private correspondence, Newton monitored the smell in minute detail. On January 3, at the beginning of the whale's exhibition in Chicago, he scrawled postscript on a letter to his son Warren, "We have had no occasion to use Ice or any thing yet to keep the Bird—."[42] On January 4, he repeats, almost as an incantation: "The critter is keeping 1st rate have not had to use any thing & if the weather continues any way cold I see no reason of any."[43] It is telling that Newton calls the whale the "Bird" and the "critter"; these are the elliptical ways in which his account marks the whale's ontological slippage through its putrefaction. The nicknames allowed Newton to mark change without admitting to himself that the whale, the linchpin of his scheme to make a fortune, was transforming into something else.

Later in the tour, the whale's growing smell became an impediment to profit, but it is an example for us of how putrefaction makes dead bodies vibrant, insistently present. After the whale was treated in Cleveland in April 1881, the show traveled to Toledo and Detroit, where on account of the smell Newton and Engelhardt were forced to leave town early. A railroad correspondent to the *Kalamazoo (Michigan) Telegraph* reported in May 1881 that he smelled the whale in Toledo without having seen him: "Toledo! Whew! What a smell! Fishy smell! To the heavens it seems to swell! We ask our friend if he is acquainted here; he says no; so t'would do no good to ask why this aroma? But passing along, a hand-bill is thrust into our hands, telling us of '*the*' whale dead and in bad odor, being in the city, and then we understood whence this all-pervading perfume."[44] As spring progressed, and the whale thawed, the radius of the whale's influence on the towns it visited expanded so that by the time the whale reached the Upper Midwest, people in whole sections of each city could unwillingly smell the whale without seeing it.

As is probably evident by now, accounts of the whale show are almost invariably humorous—punny, allusive, and winking—never more so than when addressing the whale's odor. Jokes associated with bad smells are often deemed childish and unserious, an attitude itself with deep historical roots, according to Jenner.[45] The jokey, fluky quality of the whale show makes it an awkward fit for environmental humanities scholarship, which is coalescing around an ethics of care and affective stances such as concern, melancholy, and grief—with good reason, in light of climate change, mass extinction, and other ecological crises. And, in a post-Save-the-Whales world in which the exhibition of *live* charismatic megafauna—especially whales—stirs violent controversy, it can be difficult to see the exhibition of a whale's corpse as anything other than an atrocity.[46] In that light, the humor appears grotesque. But the nineteenth-century humor surrounding the whale show is an instructive affect. The whale show humor is a social expression of anxiety—anxiety about unseemly encounters with advanced putrefaction that threaten bodily boundaries and about the very alienness of the whale itself. It is not fully true to the ethical thrust of Alaimo's theory to say that encounters with the dead Prince of Whales incited an experience of transcorporeality that could unseat human exceptionalism.

But the Prince of Whales show did not rigidly affirm human exceptionalism, either; encountering the dead whale incited a sensory experience that made human and nonhuman bodies porous. The dead whale challenged Newton's and Engelhardt's capitalist ambitions, too. Philosopher Henri Bergson theorized laughable objects as those that contain "something mechanical in something living," and it is this quality that makes Newton and Engelhardt into objects of humor.[47] The promoters acted mechanically, heedless of a putrefying body's unaccountable vitality, attempting to commodify the whale's body and earn a profit from it. The whale itself resisted objectification by changing before their very eyes.

THE HYPERMEDIACY OF THE WHALE: OBSOLESCENCE AND INFRASTRUCTURE

Hypermediacy is the second face of remediation's double logic: the quality of a remediated object that calls attention to its own artifice and to the mechanisms that make its transmission possible. The designers and consumers of the Prince of Whales show produced a prolific body of material on the means of the whale show's mediation—that is to say, the infrastructure that made its exhibition possible. The appearance of the pelagic beast in the middle of the country was not performed as a feat of magic; Engelhardt and Newton promoted the show as a spectacle of logistical mastery. What the whale show's hypermediacy reveals is the layered, sedimented qualities of transportation and energy infrastructures and the unevenness—even the violence—of infrastructural and technological change. The whale's body was delivered ashore by a wooden whaleship (a site of the old organic energy regime) and transferred to a rapidly modernizing, coal-powered railroad (a site of the new mineral energy regime). But at each moment of transfer, labor teams powered by humans and animals working together maneuvered the unlikely cargo to and from the main rail lines. Serious accidents occurred all along the way. One occurred when the whale was being installed in Cincinnati, Ohio, in February 1881: the exhibition shed built for the whale suddenly collapsed, causing "serious internal injuries" to one worker.[48] And when the whale show arrived in Philadelphia in March 1881, workers leading a team of mules tried to maneuver the railcar bearing the whale on a custom-built sidetrack near the viewing platform erected for the show. The team lost control of the railcar, and the whale crashed twice into the platform, "astonishing the mules, and scattering the broken timbers over the huge carcass and under the cars."[49] The transitions from organic to mineral energy regimes, from sail to steam, from wood to steel, did not follow a smooth teleology but erupted in a series of collisions that happened (and continues to happen) violently and in plain sight. The violent collision of old and new infrastructures represented by the whale's journey overlays the longer and deeper violence of Indigenous genocide and erasure caused by settler colonialism in the North American West. The word "pioneer" in the Pioneer Inland Whaling Association makes no effort to hide the history of settler-colonial violence, which it cloaks in the settler ideology of innovation still transmitted today by the word "pioneer."[50] The whale on a train that crashed and careened through the Midwest resembles the fearsome symbol that journalist Frank Norris offered some twenty years later in his muckraking account of the railroad industry. Given the epistemological problems posed by sea creatures, it is little surprise that Norris called his assemblage *The Octopus*, another mysterious Leviathan.[51]

The hypermediacy of the whale show—and of other remediated objects on the media-infrastructure spectrum—speaks to an ongoing debate about the visibility of infrastructure. An observation from an early text in infrastructure studies describes working infrastructures as invisible until moments of breakdown.[52] This early insight remains a valuable point of discussion in the field, even though the claim about infrastructure's invisibility has been widely repudiated: infrastructures are highly visible to people who labor at creating and maintaining them, whose lives and lands are damaged by them, and even by those who choose to see them for any number of reasons.[53] The whale show's hypermediacy—the attention it calls to its own mediated nature and transmission across space and time—is one instance in which the infrastructure of the railroad system becomes hypervisible. Two narratives of the whale show's transport from the Atlantic Ocean to Chicago—stories that Newton and Engelhardt repeated and newspapers continuously reprinted throughout the two-year tour—illustrate the show's entanglement in old and new transportation systems. Just as the examples above indicate the often violent clash of organic and mineral energy regimes, these narratives show how newfangled and old-fashioned infrastructures became enmeshed. A single traveling entity traversing old and new systems takes on the flickering associations of fossil modernity and obsolescence.

In the first narrative, the Prince of Whales serves as an illustration in the operation of railroads and as a limit case in the railroad's modern capacity to deliver almost anything almost anywhere. The shipment of the whale was published in one newspaper as an episode of "Rail News." The newspaper reports the whale's shipment in the form of a dialogue between Engelhardt and a deadpan rail agent, who is unflapped by the promoter's request to ship a whale.

"You are on the right track this time, young man," was [the agent's] reply. "We make a specialty of whales. Let me see. Here is our tariff. Whales, I believe, come under the heading of W—, W—, ah! Here it is—whales—whales—well now, this beats me; I was real sure we had whales in our tariff, but somehow I can't find it. Last week we shipped an elephant, and only last Saturday eight sea lions, and being in the menagerie business, you know, I took it for granted that the fellows who arranged this tariff sheet had included whales, but I guess I am mistaken. This thing must be looked after. How big is your fish anyway?"

Matters were then explained. . . . What Mr. Engelhardt wanted was permission to allow the whale to remain on the cars on arrival here, and during a stay of three weeks, while it is exhibited in the Exhibition building. Permission was granted, a track will be constructed into the building, and there can be seen what is undoubtedly the biggest thing that was ever seen on wheels.[54]

The humor in the passage centers on the railroad clerk's nonchalance when confronted with the task of assigning value to the shipment of the whale. Although the line about railroads being "in the menagerie business" played for a laugh, it was also true. Inland infrastructures had transported circuses for many years before the Prince of Whales traveled to Chicago. The first rail circus in the United States set out in 1856, and by 1880, the railroad was transporting whole herds of elephants.[55] In an environment of one-upmanship, though, the whale makes a competitive bid for the "biggest thing that was ever seen on wheels." Railroad development was thought to follow the logic of progress: the railroad will grow ever bigger, more powerful, ever increasing its carrying capacity.[56] In this sense, the whale marked the modernity of the railroad infrastructure by demonstrating its previously unimagined power and capacity.

The second narrative of the whale's transport demonstrates the whale's enmeshment in the maritime infrastructure. This narrative illustrates how the whale was brought to shore by an elaborate, jury-rigged process.

An immense cradle, built of Georgia pine and mortised and bolted with iron, was sunk in the dock underneath the fish, and engines set to work to pump out the dock. When the latter was dry, the whale lay on the timbers, reclining on his back, as all good and dead whales should do. He was then lashed to the cradle, care being taken to place canvas beneath the ropes so as to prevent the cutting of the flesh. The cradle and its baby occupant were then towed across the channel to the railroad dock, where, with the aid of three large derricks, he was hoisted on a couple of platform cars.[57]

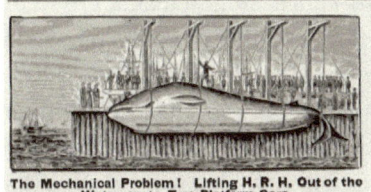
The Mechanical Problem! Lifting H. R. H. Out of the Water on to Two Platform Cars.

"Special Whale Express." The Way H. R. H. Travels.

"Nothing in fiction can compare with this reality from the realms of the Wonders of Nature."

"An incomparable lesson in Natural History for the old as well as young.

"Camp Baleine." Where H. R. H. Was Preserved to the World.

This is the first, the only one, and probably the last grand specimen of the real "right Whale" you will have the opportunity to see in a lifetime.

For Dates and Location See Newspaper Advertisements.

Fred Engelhardt created this promotional material, which survives only in fragment, to dramatize the logistics of transporting and exhibiting a whale far from the ocean. © Courtesy Mystic Seaport Museum, Collections Research Center.

The account continues at length in the same vein, offering detail after detail of whale transport logistics. With its painstaking attention to the details of labor, this account conforms neatly to the genre of maritime narrative, which from the eighteenth century through the present is marked by its technical detail and jargon.[58] Newton and Engelhardt made their own self-conscious bid for inclusion in the realm of maritime culture by naming their enterprise the Pioneer Inland Whaling Association. Newton even dubbed himself "Captain" of the enterprise. In this account of the whale's transportation, the two proprietors are the resourceful men who display a sailorly practicality—not by the work of their own bodies on a ship deck but as men with the practical knowledge to deploy the infrastructure of coastal dockyards, intercity railroads, and urban exhibition halls. This account of the whale show infrastructure demonstrates how the whale belonged to the maritime past.

THE WHALER CALLED *Progress*: A STORY

Like the whale at the center of the Prince of Whales show, the whaleship *Progress* had a long oceangoing life before its trip to the center of the continent.[59] Built at a shipyard in Westerly, Rhode Island, the ship was launched in 1843 and sailed under the name *Charles Phelps* in the Pacific whale fishery. During the Civil War, the ship was requisitioned by the Union Navy and earmarked for the so-called Stone Fleet, a group of ships sunk in Charleston Harbor in an effort to build a natural blockade of the port. But at the last minute the ship was taken over and used as a naval shore ship instead. After the war, the ship was made into a one-time coal freighter, loaded with coal, and sailed north with its shipment to New Bedford. There, the ship was once again used as a whaler, rechristened the *Progress*, and dispatched to the whale fishery of the Pacific Arctic. In the disastrous winter of 1871, when many whalers were crushed in Pacific ice, the *Progress* served as a life-saving vessel. According to accounts, the ship found open water and rescued 226 of the people whose ships had been destroyed in the Arctic ice. By the 1880s, the *Progress*, like so many other whalers, lay empty and idle at the dock in New Bedford.[60]

The second act of the *Progress* began with a group of promoters who saw a business opportunity in the inland exhibition of an artifact from the whale fishery. In May 1892, the shipowners in New Bedford sold the *Progress* to the Chicago coal magnate Henry E. Weaver and a group of Chicago businessmen who would share the opportunity and risk of the ship's inland voyage. The group of Chicago shipowners attracted press attention as they traveled to New Bedford in the spring of 1892 to purchase the ship. In an interview with the *Daily Inter Ocean* newspaper, one member of the Chicago syndicate, Ossian Guthrie, boasted about the *Progress* as if he was an old salt: "The vessel is the best whaling vessel I ever saw. It is not a tub and not a clipper, but a genuine merchant whaler.... There is not a rotten timber in her."[61] Guthrie, like Weaver, was a man of fossil modernity: he had built the first steam-powered tugboat to be used in Chicago waterways, and he was known as the "father of Chicago's drainage system" for his work on dredging the city's rivers and canals.[62] The *Progress* would join the Chicago World's Columbian Exposition as both an artifact and a museum: it would stand in as a representative of the obsolescing whaling industry and as a museum chock-full of natural history and cultural artifacts gathered from the Arctic and other whaling grounds.

The *Progress* departed New Bedford in early June 1892 under tow by a tugboat called the *Right Arm*.[63] The ship was captained by veteran whaling captain Daniel W. Gifford, whose great-great-grandson, a public historian also named Daniel Gifford, has written a comprehensive account of the history of the *Progress* and its exhibition in Chicago.[64] The ship was towed north and west through the Saint Lawrence River, where it made the uncanny transition from saltwater to fresh. At Quebec, the vessel threw off its ballast, lightening up so that it could proceed through the river and into the system of canals that connect the Saint Lawrence with the Great Lakes.[65] When the ship stopped at Racine, Wisconsin, on July 21, the crew that had managed the ship during its voyage from New Bedford was discharged. The crew had expected a longer employment, and they protested the dismissal, according to a report dramatized in the Chicago shipping news:

> It came near being a mutiny on board the old whaleship *Progress* when the crew of eight men, composed of Portuguese and Spaniards, was discharged. It was warm for a time, as the men claimed they had signed articles of agreement for eighteen months' service before leaving New Bedford, Mass. They flatly refused to leave the ship until the Captain showed them that the articles signed were for no definite time. Some of the men refused even then to take the money, and are consulting with an attorney to see if they cannot compel the *Progress* to keep them.[66]

The dramatic near-mutiny passed, and the newspapers ignored the labor dispute in favor of more celebratory news of the ship's stopover in Milwaukee.

On July 27, 1892, the *Progress* arrived in Chicago "with all her sails set and all her flags and streamers flying in the breeze."[67] The Chicago shipowners chartered a local steamer and sailed out to meet the vessel as it came into the city. Among the welcoming party were Henry Weaver, Ossian Guthrie, and the ship's one-time captain James Dowden. Nearby steam yachts fired cannons in salute as the welcoming party boarded the ship.[68] After the gala arrival of the *Progress*, it opened to visitors in downtown Chicago before the official start of the Columbian Exposition. Chicagoans came in droves to see the ship and the artifacts that were crammed into every available space on board the ship: "Trophies are everywhere, on the deck, in hold, and in the cabin, [c]ases of bones, mammoth turtles, sharks' jaws, skeletons of whales and other interesting parts, sword fish, and suggestively the ceiling is studded with stuffed star fish. There are harpoons and 'bum' guns for shooting whales and fine Fayal work made by the Portuguese women of the Azores islands from the fiber of the century plant."[69] But one month into its downtown tenure, the *Progress* met with disaster. On September 24, a dredging vessel working in the Chicago River rammed the *Progress* while schoolchildren were aboard, and the ship sank to the bottom. Many artifacts were lost in the filth of the Chicago River, but within a few days, the ship was patched up, raised, and reopened for business.

The *Progress* fared only slightly better during the Columbian Exposition, where it was exhibited in the South Pond near Jackson Park. The ship and its Arctic museum were mentioned occasionally in press coverage of the exposition, likely because the ship's high profile made it visible from different points of the fairground. The press coverage portrayed the

The *Progress* was a fish out of water at the Columbian Exposition in Chicago, an old-fashioned exhibition by which visitors measured modernity at the exposition. Albumen print, 1893. Courtesy of the New Bedford Whaling Museum.

Progress as a retrograde outlier in a fair oriented toward the future. An account of the fair by visitor Mrs. Mark Stevens transmitted her palpable boredom: "Below, the space . . . had been transformed into a museum . . . containing curiosities numbering into the thousands. We have not the slightest desire to name many of them, nor the reader to hear them, so, on mentioning a few, will pass on to something else."[70] A retrospective account of the *Progress* at the Columbian Exposition characterized the ship's journey inland and its exhibition as "a losing financial venture from the start" and reported that the owners had lost $20,000.[71] The exhibition would be remembered as a failure: "From the outset of the venture it seems they [the backers] might always as well have dumped the money in the sea, for the enterprise proved a most disastrous one."[72]

After the close of the fair, the *Progress* languished in South Pond, splintering and drifting toward the bottom.[73] The ship owners sold the collections of the Arctic museum to the Field Museum, which, after a few years, in turn sold most of those artifacts to the Peabody Museum in Salem, Massachusetts.[74] For a few years, the ship became a kind of playground for neighborhood children in Jackson Park: "The boys got into every nook and crevice of the vessel's deck and hold, and up the mains over the rotting ropes and cordage like an army of rats."[75] In June 1896, the *Progress* was towed out of South Pond and the *New York Times* reported that the ship might be sent back east for restoration and new use.[76] That and other subsequent plans proved futile, though, and the ship ended up south of the city in the Calumet River, where, in 1902, the *Progress* burned down to the water line. After the fire cleared, the ship's bottom, stuck in the mud, was broken up with dynamite.[77]

SHIP MEDIA: OBSOLESCENCE AND INFRASTRUCTURES OF SETTLER COLONIALISM

Like the Prince of Whales, the whaleship *Progress* offered its Chicago audiences the opportunity to witness the remediated ship's hypermediacy: the means of its physical transmission from saltwater sea to freshwater lake and from working vessel of extractive industry to leisure-time tourist attraction. The *Progress* was a metonym for the maritime and whale oil–based infrastructural regime of the past and—at the same time—a benchmark for the fossil-fueled infrastructural regime of the present.[78] The *Progress* was on hand at the Columbian Exposition as if to say: "Look how far we have come from *this*." It is telling that Henry Weaver, Ossian Guthrie, and the other Chicago owners of the *Progress* were men whose wealth and political prominence was founded on fossil modernity—on the very forces that were bringing the US commercial whaling industry to an end. Weaver made his fortune in the railroad and coal trade. According to historian Daniel Gifford, Weaver turned to the *Progress* scheme only after his company lost the contract for supplying the Columbian Exposition with fuel (Sinclair Oil won the contract).[79] By 1894, Weaver's coal company was the largest in Chicago; at that time, it was shipping 600,000 tons of coal into the city.[80] Ossian Guthrie literally brought steam-driven vessels to Chicago waterways and physically remade the canals and rivers of the city through new dredging and drainage technologies.[81]

The accident that sank the *Progress* in the Chicago River shortly after its arrival highlighted the ship's unsuitability to inland infrastructure and the violent clash between old and new infrastructural regimes. The *Progress* was rammed and sunk by a sand scow; a dredging barge attached to a steam-powered towboat that maintained the navigability of Chicago's artificial waterways. One newspaper report even suggested that the collision was a deliberate strategy on the part of impatient tugboat sailors to get the old ship out of the way: "The scow that ran into the *Progress* was in tow of the tug Munson of the Vessel Owners' Towing Company. The expressions of satisfaction that were made by vessel men when they heard of the sinking of the boat led to the report that the collision was intentional. The *Progress* has been in the way of tugmen ever since she was tied up at the State St bridge."[82] Other reports figuratively, if not literally, cheered the accident by highlighting the contrast between the banality of the accident and the dignity of the ship's perilous past: "A Chicago sand scow sunk the gallant vessel that had laughed at danger for so many years."[83]

The *Progress* was a fish out of water, too, at the future-oriented world's fair. Press and visitor accounts unanimously stressed the ship's startling antiquity. The adjectives "old" or "ancient" are attached to nearly every account of the ship before, during, and after its exhibition at the exposition. The ship's former captain, James Dowden, who served as a kind of tour guide aboard the vessel, was unfailingly called an "ancient mariner" and was treated by the press as a vestige on par with the whaling "relics" on exhibition.[84] Historian Daniel Gifford notes perceptively that the "Complete Marine and Whaling Museum" on board the *Progress* was deeply out of step with contemporary museum practices. According to Gifford, the museum on board the *Progress* bucked the Gilded Age trend for "classification and organization" in favor of an anachronistic hodgepodge style of presentation more like P. T. Barnum's museums of an earlier generation.[85] While the madcap Midway and the exhibitions of dynamos and steam-driven machines were characterized by their frenetic energy, the *Progress* was a site where visitors could slip into a languorous reverie of the past: "There are not many days at the Exposition when one cannot find sunning themselves on its [the ship's] broad decks a group of old sea-dogs who forget the sights of this great nineteenth-century show in recounting thrilling experiences of the period to which the *Progress* belongs."[86] Other less romantic accounts characterize the *Progress* and the stories it told as boring and irrelevant to public interest: "Some way or other fair visitors did not want to know how a whale was harpooned or how sailors tunneled into him and excavated his bones, or how his oil was boiled. The apathy which the people evinced toward the whaling scenes was remarkable."[87] The whaling ship's name heightened the irony of its failure at the exposition.

And yet, the presence of the *Progress* at the exposition was also consonant with the racial and evolutionary ideology of the fair: the past gives way to the future in the way that "savagery" gives way to "civilization"—and through the same violent means of settler colonialism and extractive capitalism. The Columbian Exposition was staged in honor of the four-hundredth anniversary of Christopher Columbus's arrival in North America; the structure of settler colonialism was the ground on which the fair's utopian vision of the American future was built. Historian Robert Rydell explains that this ideology was structured into the architecture of the exposition, which was divided roughly into two sectors: the White City and the Midway Plaisance. The White City was the utopian vision of an American future, a racially coded white place made brighter and whiter by electrical lights; the midway was the fair's wild unconscious, where the Ferris wheel and fairground games and treats abutted the anthropological exhibition of Indigenous people and artifacts. The midway, in Rydell's view, "provided visitors with ethnological, scientific sanction for the American view of the nonwhite world as barbaric and childlike and gave a scientific basis to the racial blueprint for building a utopia."[88]

The Prince of Whales simultaneously traversed old and new systems and, in so doing, took on the flickering associations of modernity and obsolescence. The same is true of the whaleship *Progress*: it connotes both obsolescence by example and modernity by contrast. The *Progress* stands in similarly conflicted relationship to the racial and colonial politics of the Columbian Exposition. The exhibition of old and new technology at the exposition were not racially neutral, and the story of the *Progress* at the world's fair demonstrates the complex racial and colonial dimension of hypermediacy. The *Progress*, resting in the South Pond near the Machinery Building, indexed the distance between past and future technology and, by the Columbian Exposition's transitive logic, the distance between savagery and civilization. The *Progress* was marked by its otherness, cast by the strange racial logic of the fair as an Indigenous or racially othered body, on exhibit in order to be performatively dominated and surpassed. The exposition performatively restaged colonization in several different forms. Rydell recounts that several Sioux chiefs were invited to highly visible seats at the exposition's opening ceremonies and made their appearance at the moment when a chorus began singing "My Country 'Tis of Thee."[89] The *Progress* was similarly on hand in order to perform its obsolescence in strangely racialized terms. The characterization of the *Progress* as "gallant" resonates with the colonial discourse of the "noble savage," while accounts that emphasize the ship's antiquity and quaintness sound like the colonial discourse of the "vanishing Indian."

At the same time, though, the *Progress* was in every way an instrument of settler colonialism in its life as a whaler and as a museum ship. Like all other ships of the US commercial whaling fleet, the *Progress* had delivered white settlers to emerging colonies around the world—and with those settlers, all of the violent accoutrements of colonialism, including war, disease, and extractive capitalism. Beginning in the nineteenth century with the US fleet, commercial whaling disrupted long-standing patterns of Indigenous whaling, especially in the Arctic, where sustainable whaling had long provided food for large communities.[90] And the US commercial whaling industry was one of the most powerful sites of colonial knowledge production through its contributions to mapping and artifact collection. The *Progress* staged its own colonial violence at the Columbian Exposition in the form of its onboard museum, which included, alongside natural history specimens and whaling implements, a navigational quadrant allegedly transported in the *Mayflower*, clothing and tools taken from Indigenous people in the Arctic and the South Pacific, and even the body of an aboriginal boy from Australia. As Mrs. Stevens wrote, he was "a mummified Australian boy, five hundred years dead; the wind blew him out of the tree in which he was fastened, and sailors brought him away."[91] The museum aboard the *Progress* neatly conflated the narrative of *Mayflower* settlers with the long-standing and global scope of whaling colonialism and extractive violence to humans and nonhumans. Daniel Gifford has written that the "higgledy-piggledy mixing of whale bones and shells with mummies and grass skirts" in the *Progress*'s onboard museum reflected a deeply anachronistic museum exhibition strategy.[92] At the same time, the undifferentiated mixture of material objects aptly represented the entanglement of settler-colonial violence with whaling and other extractive pursuits. The *Progress* was, simultaneously, a racialized specimen and a ship freighted with the victims of extractive colonialism.

As the hypermediacy of these inland whaling shows suggest, old infrastructures do not disappear; they migrate from the sphere of usefulness to a sphere of entertainment, all while flickering between strange and sometimes contradictory ideological uses. Marshall McLuhan wrote fleetingly about this migration in a chapter of *Understanding Media* on the "hot, explosive medium" of the car: "Today small children plead for a train ride as if it were a stagecoach or horse and cutter: 'Before they're gone, Daddy.'"[93] Old infrastructures are sometimes regressive, ludic media; at other times, old infrastructures work with fierce intensity to uphold contradictory political ideologies. The expanding railroad system, powered by fossil fuels, delivered pelagic whales and oceangoing vessels to inland spectators and, moving in the other direction, inland tourists to New England port towns that reoriented their economies toward tourism when the sailing trades began to disperse.[94] Obsolescence is what happens after an industry—its laborers, infrastructures, technologies, commodities—dies but does not disappear. In the body of the Prince of Whales and the whaleship *Progress*, putrefaction and obsolescence intersect, demanding that we account for the complex forms of life that proliferate after death.

PART THREE
Whaling Nostalgia

4

Extinction Burst

White Supremacy and Yankee Whaling Heritage at the End of the Industry

Around the same time that neighborhood children in Chicago climbed all over the abandoned whaling ship *Progress*, children in New Bedford, Massachusetts, played on the abandoned whaleships in their own backyards. One of those grown-up children, Clifford Warren Ashley, wrote of his childhood:

> For the fortunate youth of that day the unpoliced ships and grassgrown wharves made a marvelous playground. We learned to swim from the bob-stays of the old hulks. We contrived to paddle and row on rafts fashioned of hatch-covers, and used in boarding parties overside. We swarmed over the rigging and slid down the backstays, spun the wheels, and on rainy days gathered in the cabins and played games and pretended one thing and another; and always it was something that smacked of the sea. Often in winter we skated up the river to where an old whaler was stranded, and there built fires of sheathing ripped from her sides. In those days we didn't "shoot" marbles; we "pooned" them with an overhand toss, just as a harpo[o]neer darts his iron; and when one boy was chased by another, he "went fluking" down the street. There were whole wharves along New Bedford's waterfront that served no purpose save that of a playground, and there was never another like it.[1]

During their carefree boyhoods, Ashley and his friends disassembled the whaling infrastructure that was growing obsolete before their eyes: they broke off parts of abandoned ships and repurposed them as rafts and kindling. Later, in somber, nostalgic adulthoods, they experienced the decline of New England whaling as a profound loss, and they came to a new understanding of the wooden remnants as sacred relics to be carefully documented, protected, and preserved for the future.

Whaling culture in part 2 of this book, "Whaling Entertainment," served as a site for leisure, entertainment, and play: the obsolescing infrastructures of the industry were made into ludic sites for tourism and entertainment. The children playing on the ruined *Progress* in Chicago were cited as evidence of the whaling ship's final failure and abandonment. But the grown-up children who remembered playing on whaling ships in New Bedford took those memories far more seriously. In part 3 of this book, "Whaling Nostalgia," I trace a shift in the emotional tone and ideological work of whaling culture in the first three decades of the twentieth century. During this period, the decline of whaling became a cause for serious concern, melancholy, and hard thought about the changes wrought by fossil modernity. Whaling culture became deeply nostalgic.

In the first three decades of the twentieth century, a cohort of civic leaders, artists, writers, historians, and filmmakers in New Bedford worked together to document and preserve the history of the US whaling industry. During this period, fewer and fewer whaling voyages launched from New Bedford, and this cohort was keenly aware that they were witnessing the end of something. These men included local political leaders, such as William Crapo, a lawyer and congressman from New Bedford who was a founding member and first president of the Old Dartmouth Historical Society. Others were wealthy heirs, such as Edward Howland Robinson Green, whose mother, Hetty Green, invested her family's whaling wealth and became known as the notorious Witch of Wall Street. Some of these men, such as the artists and historians Albert Cook Church and Clifford Warren Ashley, had grown up in New Bedford at the end of the nineteenth century and cherished boyhood memories of playing in the obsolescing port.

This chapter tells the story of four of this cohort's preservation efforts: a civic monument, an essay, a major Hollywood film, and a whaling ship turned into a museum. New Bedford preservationists did much more than I am able to document here, and they did so for many different reasons and to many different ideological ends than those I describe here. There were other preservationists who collected artifacts and documents and established the whaling museum; who founded the local historical society; who built a half-scale model whaling ship. But I focus on a particular cohort of individuals and collectives whose ideas, feelings, and forms are distinctly nostalgic. In working to preserve their local history, this New Bedford cohort was not exceptional; their efforts were representative of a wide movement of revivalism and cultural institution building in the

United States in the decades around the turn of the twentieth century.[2] But the institutionalization of whaling history in the early twentieth century in New Bedford cannot reliably be called a revival because the New Bedford whaling industry, though dramatically diminished from its mid-nineteenth-century peak, was still in operation. What civic leaders in the early twentieth century put on for the city through these institutions was the observance of a passing: in short, a funeral.

The specific group of early twentieth-century whaling preservationists chronicled here went to extreme efforts to characterize the whaling industry that they mourned as white American heritage. According to the narrative that emerged in some of these preservation efforts, the passing of so-called Yankee whaling represented a larger threat to whiteness itself. Whaling heritage in this place and time served a form of white supremacy shaped by nativism, the race science that culminated in eugenics, and an antimodern impulse that associated artisanal labor with masculine virility. Given that the whaling industry of past and present was a space of enormous racial diversity, this was no easy task, and the history of whaling preservation in New Bedford as told here is the history of herculean feats of historical misinterpretation and strenuous whitewashing. The process of creating a white Yankee whaling heritage oscillated between extensive research and refusal of the facts. This chapter measures the distance between the evidence available to these whaling heritage makers and the stories that they ultimately told. A particular understanding of whiteness, one defined by anxieties about decline, degeneration, and even extinction, came to be attached through the practices of heritage to the obsolescing whaling industry. The case of Yankee whaling heritage helps reveal how stories of energy and technological obsolescence can be coopted into the specious declension narratives of white aggrievement and white supremacy.

The somber, funerary efforts of early twentieth-century whaling heritage makers is an instance of what I call extractivist nostalgia, following anthropologist Renato Rosaldo, who coined the term "imperialist nostalgia." Rosaldo described imperialist nostalgia as the way that imperialist colonizers "long for the very forms of life they intentionally altered or destroyed" and "mourn the passing of what they themselves have transformed."[3] The very same people who sought to preserve whaling heritage were those who benefitted from the economic transformations that were ending wooden-ship commercial whaling in the Northeast: those who had grown wealthy in and through fossil modernity. And yet these were the preservationists who felt so strongly and so sadly about the end of New Bedford whaling that they grieved it as a death. Extractivist nostalgia led to preservation attempts that underscored the whaling industry's role in US imperialism, white supremacy, and patriarchal masculinity.

I center race and whiteness as the main category of analysis here, but it is impossible to separate whiteness from masculinity, especially in whaling and other energy discourses.[4] The white masculinity I discuss is a forerunner of what Cara Daggett has called "petro-masculinity": a concept that explains how fossil fuels and white patriarchal order "are mutually constituted, with gender anxiety slithering alongside climate anxiety, and misogynist violence sometimes exploding as fossil violence."[5] Whiteness and masculinity are co-constituting ideologies in the early twentieth century, and although my study of Yankee whaling heritage focuses on its white supremacy, the patriarchal masculinity of the movement is inextricably woven into it. This extractivist nostalgia continues in the present day in the nostalgia of fossil fuel companies, climate change deniers, and local communities for the US coal industry that coalesced during the Donald Trump administration as a campaign to save coal—and through coal, whiteness itself.

FIRSTING, LASTING, AND EXTINCTION

The case of early twentieth-century Yankee whaling heritage reveals the logics of white supremacy and settler colonialism at work in specific ways at moments of energy transition and industrial change. I am far from the first scholar to observe the link between natural resource extraction and white supremacy; the field owes the insight primarily to Indigenous scholars and activists who have documented and demonstrated the extent to which natural resource extraction is part and parcel of imperial violence and settler colonialism.[6] This reading of Yankee whaling heritage builds on the foundational work of Jean M. O'Brien, who showed how nineteenth-century New England historical practices bolstered settler colonialism and white supremacy through the practice she called "firsting and lasting": compulsively repeated stories about the first settler or last Indigenous person in a particular New England community. "Lasting" is a narrative strategy that allowed nonnative New Englanders to cast Indigenous people into extinction, even though Indigenous people have lived continuously in New England.[7] It might seem surprising or counterintuitive on the part of local commemorators to apply to the whaling industry the same strategy of "lasting" that accorded to Indigenous people, as if heritage makers were "playing Indian" themselves.[8] But O'Brien's account shows that nineteenth-century New England histories were rife with other "lasts," too: accounts of the "last" deer of Rhode Island, the last fishery in Topsfield, the last Tory in Connecticut, the last surviving children of the first settlers—and, as this chapter chronicles, the last Yankee whaling ships.[9] These non-Indian "lasts" do not depart from the larger anti-Indigenous project of lasting; they cement the idea that modernity is born through the extinction (that is, the extermination) of the old. "Lasting" is a practice that seemingly follows the affects and forms of grief, but it does so in service of consolidating power in the hands of those that have brought about the loss.

The very concept of extinction grew up around and within these stories about threats to whiteness, and extinction itself acquires special potency in Yankee whaling heritage, which celebrated an industry that always has been what John Levi Barnard has called an "extinction-producing economy."[10] The extinction being mourned by the Yankee whaling heritage makers was the supposed "extinction" of the Yankee whaling industry and its quaint artifacts and customs: not the whales under direct threat of extinction by the modernizing whaling industry that would exponentially ramp up its scale of slaughter, nor the coming extinction of untold species in the Sixth Great Extinction, nor Indigenous people live under the conditions of continual genocide, nor the people of the world who experience the existential threat of slow violence under extreme climate change. But nevertheless, "extinction" became a powerful discourse in Yankee whaling heritage. The commemoration of the Yankee whaling industry was a form of grief about extinction, but the objects of that grief (dilapidated whaling ships and implements and idealized fictions about the "type" of man who worked in the industry) work within a very different political and emotional matrix than the type of extinction grief necessitated by climate change in the twenty-first century. And yet the rhetorical structure of extinction grief then and now is similar. The old white veterans of the nineteenth-century whaling industry and the artifacts and infrastructures of whaling themselves were treated as "endlings," the last of their species.[11] It is because it served whiteness that so much of early twentieth-century Yankee whaling heritage took the form of a funeral: the obsolete Yankee whaling industry evoked the specter of race extinction that hung over early twentieth-century white supremacy. The extinction of the white race always was a paranoid fantasy, but it was a powerful fantasy that diverted the emotional and political work of apprehending extinction away from actual threats.

THE INVENTION OF YANKEE WHALING

In the process of holding a funeral for a dead industry, the New Bedford civic leaders had to invent a corpse. They called it "Yankee whaling." The industry that civic leaders invented and commemorated in the early twentieth century was a very particular form of the industry: whaling undertaken for the production of specific commodities (oil and, to some extent, whalebone), carried out through specific technologies (hand-thrown harpoons from small wooden whaling boats), supported by specific infrastructures (wooden sailing ships), financed by a specific financial structure (with individual voyages financed by a group of investors and shipboard laborers compensated through the "lay" system unique to whaling), and launched from New England port cities (especially New Bedford). And the feature of Yankee whaling perhaps most sacred to these early twentieth-century commemorators was the racial identity of the laborers who prosecuted the industry. They were known in the contemporary parlance as Yankees, "native" Americans, "Pilgrims," "Quakers," "hardy" and "God-fearing" men, and men of the "old stock" or "old type": all of which is to say that the old whalers that were commemorated in this period were white men descended from English settlers.

"Yankee whaling" was not a complete fabrication, nor were the New Bedford civic leaders wrong to say that a form of the industry was dying: many whaling voyages in the nineteenth and early twentieth centuries resembled the archetypal "Yankee" voyages as described above, at least in part. And, indeed, the whaling industry was radically shrinking and the wooden sailing ships that once dominated local harbors were falling into disuse. But the financial structures, technologies, and infrastructures of the whaling industry were not nearly so homogenous as the Yankee whaling heritage makers claimed, nor was whaling coming to any kind of end; the global whaling industry would slaughter more whales in the twentieth century than ever before in history.[12] But the biggest distortion made by the commemorators of Yankee whaling had to do with the identities of the industry's laborers. The crews of whaling voyages were and never had been all white, Anglo-American men. "Yankee whaling" as commemorated in the early twentieth century was a whitewashed fiction. In actuality, laborers in the whaling industry aboard ship and on shore were stunningly diverse and included large communities of dark-skinned Portuguese immigrants from the Cape Verde islands, Indigenous people from New England nations, Black Americans, South Pacific Islanders, and many others.[13] In fact, as the industry declined, people of color—notably Cape Verdean laborers—made up an increasing proportion of the whaling labor between 1900 and 1925. Though early twentieth-century histories of whaling grudgingly admitted the presence of Black and brown Portuguese sailors in the dwindling whaling industry, they perpetuated the myth that Cape Verdean sailors—often called "Bravas," after the Cape Verde island—were only the most lowly of laborers. In fact, more and more shipowners and vessel masters were Portuguese, and increasingly, the language spoken aboard ships and in which New Bedford ships' logs were recorded was Portuguese, too.¹⁴ White whaling historians whitewashed the industry's past, and they characterized the changing demography of the early twentieth-century industry with undisguised racism. Elmo Hohman in *The American Whaleman* even attributed the industry's downfall to its diversifying labor force:

> And as the better types of Americans forsook the forecastles, their bunks were filled by criminal or lascivious adventurers, by a motley collection of South Sea Islanders known as Kanakas, by cross-breed negroes and Portuguese from the Azores and the Cape Verdes, and by the outcasts and renegades from all the merchant service of both the Old World and the New. This extraordinary mixture of races, nationalities, and types had also characterized the fishery, it is true, throughout the [eighteen] forties and fifties. But after 1865 the dilution of the labor force by an ever-mounting percentage of ignorance, incompetence, and general inefficiency proceeded at a still swifter pace. . . . This progressive deterioration in the character, skill, and efficiency of the crews lay like a rock in the path of whaling success.¹⁵

Hohman's outright racism is characteristic and instantly recognizable as such, particularly when he uses such terms as "dilution" and "deterioration" to describe the racial diversity of the whaling labor force. Other forms of white supremacy in whaling historical preservation are more insidious.

The efforts of early twentieth-century New Bedford civic leaders are best understood as heritage rather than history, to draw upon a distinction made by historian and geographer David Lowenthal. Heritage is a historical practice meant to join and serve the members of a community, to celebrate the origins and justify the stature of an exclusive group. As Lowenthal explains, "History tells all who will listen what has happened and how things came to be as they are. Heritage passes on exclusive myths of origin and continuance, endowing a select group with prestige and common purpose."¹⁶ The creation of Yankee whaling heritage in New Bedford in the early twentieth century misread the wider history of the industry; it excluded and even erased the nonwhite people who worked in the industry, and it celebrated violent resource extraction and exploitative labor practices. More interesting than pointing out the various historical errors and distortions of this particular form of heritage is understanding the interests that this heritage served. At the most basic level, Yankee whaling heritage served the interests of the elite white New Englanders who carried it out: men whose wealth was built on violent and exploitative industries such as whaling and who, unlike the laborers in the industry, did not incur its harshest costs.[17] But the scope of Yankee whaling heritage was much wider: the particular form of whaling heritage that emerged in New Bedford in the first decades of the twentieth century served the interests and ideologies of whiteness and white masculinity.

The great irony of whiteness in the early twentieth century—an irony that persists to the present day—is the way in which it was expressed as an endangered category, an identity subject to decline, degeneration, and extinction. Whiteness comes into being through stories of decline even as white people hoard power and violently suppress people of other races using every available means. White men told themselves that whiteness was in danger of dying out through an enormous range of cultural and political forms at the turn of the twentieth century: through the passage of violent anti-immigrant laws such as the Chinese Exclusion Act (1882) and the Alien Contract Labor Laws (1885 and 1887); through long-standing antimiscegenation laws affirmed in the Supreme Court case *Pace v. Alabama* (1883); through white supremacist tracts such as Madison Grant's *Passing of the Great Race* that raised alarms about race suicide and the end of the white race; and through horrific eugenic practices including eugenic euthanasia and the eugenics-inflected Immigration Act of 1924. White people expressed racial grievance and a spurious fear of racial decline more subtly but no less firmly through movements of conservation and preservation: wilderness conservation movements that served the leisure of white settlers and historic preservation practices that enshrined white history and marginalized all others. But in working to protect whiteness, whaling preservationists had to invent their subject.

THE WHALEMAN STATUE

On June 20, 1913, New Bedford leaders brought into being the prototypical Yankee whaleman: a monumental statue of a harpooner erected in the city's center in front of the Free Public Library. There the monument still stands, just a few blocks uphill from the port. The monument consists of a square granite pedestal on top of which rests a bronze whaling vignette: the forward quarter of a whaleboat riding on ocean waves and, in the boat's prow, a harpooneer standing with a harpoon raised over his right shoulder and ready to dart it into the back of a whale. The whaleman is muscular and shirtless, frozen in a moment of pure potentiality. The whaleman's fine, flowing hair (a racial characteristic that makes him pointedly white) is parted in the center of his head and echoes the parted ocean wave at the bottom of the bronze tableau; the visual rhyme between hair and water naturalizes the affinity between the white man and the

The whaleman statue was installed in 1913 in front of the New Bedford Public Library, where it still stands today as a figure of idealized white masculinity in the whaling industry. Courtesy of the New Bedford Whaling Museum.

ocean. Another large granite tablet frames the bronze tableau like a stone backdrop, engraved with the stylized outlines of two seagulls and a motto taken from Melville: "A Dead Whale or a Stove Boat."

William Crapo, the president of the Old Dartmouth Historical Society, proposed the statue in 1912 as a "memorial in honor of the whalemen whose skill, hardihood, and daring brought fame and fortune to New Bedford."[18] City leaders enthusiastically accepted the proposed monument, and Crapo commissioned the sculptor Bela L. Pratt to make the figure. There was no question as to its proper subject: the harpooner, the man whose first violent strike initiates the whale's death. A contemporary *Morning Mercury* newspaper account of the statue's design quotes *Moby-Dick* in order to demonstrate just how natural it was that the statue featured the harpooner: "There could be no doubt in any mind regarding the subject of the design. 'It is the harpooner that makes the voyage.' It is the harpooner who performs the task with the responsibility and the task with the thrill."[19] But it was the act of resource extraction at its most direct and violent—the moment when the man harpooned the whale—that defined the industry for this signal monumental representation and defined whiteness through the violence of energy extraction.

But although Melville and his line, "It is the harpooner that makes the voyage," were invoked frequently in the plans for the whaleman statue, the planners and sculptor departed sharply from *Moby-Dick* when deciding on the racial identity of the whaleman to be cast in bronze. Melville's harpooners in *Moby-Dick* were Indigenous and Black people: the South Sea Islander Queequeg from the fictional island Kokovoko, the Gay Head Indian Tashtego, and the African Daggoo. And the harpooners in the whaling industry of the early twentieth century in New Bedford were, for the most part, Portuguese. The planners of the New Bedford whaleman statue, however, were intent upon making the honored harpooner white:

> But [Queequeg, Daggoo, and Tashtego] are not the typical of the glorious host of whalemen who made the fame of New Bedford, valorous, hardy, God-fearing men.
>
> The whalers of yesteryear, whom the sculptor honors and perpetuates, is the Native born—"A health to the Native born, Stand up!"—young men athirst for gain and glory in the fishery, "stalwart fellows who have felled forests and now seek to drop the axe and snatch the whale lance." The time was when the boys of New Bedford were fired by the deeds of the fathers and aspired to be captains and heroes. This is the figure of the youth who stands at the prow of the boat—looking forward.[20]

It had clearly pained the founders of the whaleman statue to depart from *Moby-Dick*, so they shifted their discourse away from Queequeg, Daggoo, and Tashtego and sought a model "American whaleman of the Captain Ahab type," oblivious to the irony of celebrating a "type" of man whose specimen was a fictional character—and, at that, a murderous monomaniacal character whose petty vengeance destroyed the lives of his crew.

The white man honored by the whaleman statue and other early twentieth-century whaling commemorations was rarely called "white," but his whiteness was marked by a familiar set of designations—"Yankee," "hardy," "God-fearing," and "Native"—and by imperious claims to New England's past and future. The white settler's claim to "nativity" in the Yankee whaling discourse is an act of Indigenous erasure: an act of exclusion doubly remarkable in the history of the US whaling industry, whose laborers included so many American Indians and Indigenous people from around the world.[21] But in Yankee whaling heritage, the ideal Yankee whalemen were "boys of New Bedford" whose claim to privilege stretched backward toward early colonial history and, thence, toward England. The ideal white Yankee whaleman was also defined by his postwhaling future; he was a man with social, economic, and political mobility and ratified as the ideal Yankee whaleman in retrospect by accomplishments that took him far from his youthful station as an ordinary maritime laborer. The "true" Yankee whaleman was the boy who threw the harpoon in his youth and later ascended to positions of civic leadership and wealth open only at the time to white men. At the dedication of the statue, Crapo described the career of New Bedford's "first mayor" as the exemplary trajectory of the Yankee whaleman his statue was meant to honor: "He had stood at the masthead, in the boat as harpooner, he had 'struck the whale,' as the phrase went, and he earned the position and title of ship captain. For five years he ably filled the office of chief executive of the city."[22] Many different "types" of men threw the harpoon in the industry, but the true Yankee whaleman at the center of commemorative efforts in New Bedford could be seen only in hindsight after he had claimed positions of privilege that his whiteness made available to him.

The planners' insistence on making the harpooner in the statue white created difficulties for the sculptor. Crapo asked a local whaling agent if he could find a working harpooner (or "boatsteerer," as harpooners were sometimes called by whalemen) to act as a model for Pratt. The agent brought forward a man who was "a native of the Cape de Verde Islands," likely a Black man, who was summarily rejected. The Old Dartmouth Historical Society published without attribution an account of the artist's search for a model: "The whaleman of the statue, however, was to typify the early Yankee courage that sent New Bedford's sailors across all the oceans of the world, spearing cetaceans for oil,—so the outfitters were asked to find a boatsteerer of the old type,—the type made famous in 'Moby Dick' and the other stories of the sea."[23] The discursive workings of white supremacy are made clear here, where racial "type" and the facts of professional activity were conflated with such personal characteristics as "courage." The working Cape Verdean whaleman brought to the sculptor had in fact gone "across all the oceans of the world, spearing cetaceans for oil," but in the eyes of those building a monument, only a white man could typify "Yankee courage." Crapo and the agent eventually found a white New Bedford man named Richard Lewis McLachlan. They took McLachlan to the Old Dartmouth Historical Society, where they posed him in a whaleboat holding a harpoon from the museum's collection. They photographed McLachlan for the sculptor who reviewed the shots from his studio in Boston before inviting the man to meet him there. A five minutes' walk downhill from the public library would have brought Crapo, the whaling agent, and the sculptor to New Bedford's port, where, though whaling voyages were increasingly rare, they would have found a working harpooner, whaleboat, and harpoon. But the Yankee whaleman of the heritage makers' ideal had to be invented.

OLD SPORT: SPORT HUNTING AND YANKEE WHALING

With his harpoon held overhead by his strong arm, Pratt's bronze harpooner is pictured as a valiant hunter at the apogee of the chase. As such, the whaleman statue was emblematic of a widespread representational strategy that cast the bygone Yankee whaleman as a sport hunter—and at a historical moment when sport hunting was the site of enormous symbolic and real political power in the United States. The implicit and explicit comparisons between whaling and sport hunting facilitated the task of harnessing whiteness to a strikingly diverse US whaling industry. The whalemen-as-sport-hunters analogy also brought Yankee whaling heritage into relationship with the budding US wilderness conservation movement and its coeval movement, white supremacist eugenics.

"Sport hunters" defined themselves in contrast to market hunters, who sold their game for money. What made hunting "sporting" was an elaborate code of ethical practices that sport hunters touted (but did not always follow): the sportsman hunted for sport, not money; the sportsman killed only adult males; the sportsman hunted animals "in fair chase," which means that they gave animals a fighting chance to escape; and the sportsman did not hunt a species to the point of extinction.[24] But in reality, the distinction between sport and market hunters was, as historian Dorceta Taylor explains, "fuzzy." Sportsmen resorted to all sort of herding and trapping practices in order to kill animals en masse, and sport hunters often sold their game and trophies for profit. And sport hunters including Theodore Roosevelt indeed often boasted about killing the last of an animal's kind and bringing about its extinction.[25] Environmental, "sportsmanlike" rules in hunting became a way for white elite men to maintain class and race hierarchies in hunting and establish distance between wealthy white men and all others.

Sport hunting in the United States was an exercise of power, a performance of domination: of humans over nonhumans, of colonial settlers over Indigenous people, of principled sport hunters over mercenary market hunters, of masculinist and ableist white men over feminized and enfeebled others. In the United States, so-called sport hunting was practiced and governed by a very small group of elite white men, mostly from eastern cities: hunting was an affectation of wealth, one of the many ways that US capitalists sought status by parroting the habits of British aristocrats.[26] Sport hunting lay at the heart of what historians have called the "cult of true manhood" that emerged in the late nineteenth century and held that masculinity identity was forged through violent experience and physical challenge.[27] True manhood and settler colonialism mutually constituted each other on hunting trips in the US West. Hunting expeditions contributed to Indigenous genocide and the settlement of the West as hunters murdered Indigenous people and violently seized land and animals. And hunting expeditions also served as a ritual performance of settler colonialism: each hunting expedition was a miniature reenactment of Manifest Destiny, an intimate encounter with the land, animals, and natural resources that, to the hunters, seemed preordained to the use and stewardship of white settlers.[28]

Sport hunting culture accrued new prominence with the ascent of famous hunter Theodore Roosevelt to political power and, in 1901, to the US presidency. Roosevelt brought with him a cohort of political cronies and hunting friends from the Boone and Crockett Club, which he had helped establish in 1888 in order "to promote manly sport with the rifle."[29] Boone and Crockett Club members undertook wildlife preservation and land conservation as part of the project of promoting sport hunting; it did not seem paradoxical at the time for hunters to work to preserve wild animals, if for no other reason than that they wanted to sustain animal populations for future hunting expeditions. A cohort of Boone and Crockett Club members including Madison Grant, George Bird Grinnell, and Gifford Pinchot institutionalized wildlife management and conservation by forming organizations including the New York Zoological Society, the American Bison Society, the National Audubon Society, the American Game Protective Association, and others in the first decades of the twentieth century.[30] Under Theodore Roosevelt's administration and owing in large part to the advocacy groups founded by Boone and Crockett Club cronies, the first wildlife conservation laws were passed.[31]

Sport hunting as an ideology and elite sport hunters in the Boone and Crockett Club also promoted another violent practice that they peddled under the name of conservation: white supremacist eugenics. Hunter and wildlife conservationist Madison Grant directly applied the theories of species survival in wildlife management to human races who he thought should be similarly managed through eugenic practices such as immigration restriction, antimiscegenation laws, and even forced sterilization. Grant's book *The Passing of the Great Race* (1916) argued that that blond-haired, blue-eyed, "fine"-featured Nordic people were the superior race of humanity and were under dire threat of losing their superiority through threats to genetic purity such as immigration and miscegenation.[32] The Third Reich in Germany widely applied Grant's theories in devising their program of forced sterilization and mass genocide of Jews and other "social undesirables."[33] Conservation and this particularly violent strain of eugenic white supremacy both trace their origins to the social milieu of sport hunters and the fraught ideal of sportsmanship. Environmentalism throughout the twentieth and twenty-first centuries has been haunted by the legacy of violent white supremacy in these early days of conservation, and environmentalist efforts over the years have often had the effect (intended and unintended) of reinforcing class and race segregation and the displacement and erasure of Indigenous people.

With the intertwined histories of sport hunting, wildlife conservation, white supremacy, and eugenics in mind, I offer an alternate definition of "sport" in the early twentieth century: violence cloaked in the language of conservation. Environmental historians and activists have begun to excavate the legacy of white supremacy in wildlife conservation, and sport hunting has emerged in this history as one of the practices that brought violent white supremacy and conservation efforts together.[34] I build on these historians' work to show how the cultural forms of sport hunting traveled to other realms of cultural production—namely, to Yankee whaling heritage in New Bedford. There, too, as in TR's cult of true manhood and Madison Grant's eugenics, Yankee whaling in the guise of sport hunting was celebrated not in spite of, but because of, its bloody violence.

Yankee whaling heritage makers in the early twentieth century portrayed nineteenth-century Yankee whaling as sport hunting par excellence. In order to carry out any part of the analogy between sport hunting and whaling, Yankee heritage makers observed a stark contrast between the old-fashioned Yankee whaling they revered and the newfangled "modern whaling" that made it obsolete. "Modern whaling" was often called "Norwegian" or "British" whaling since those countries overtook the United States in whaling dominance in the early twentieth century. Periodizing industrial whaling into different phases, and differentiating different practices of whaling from one another is, in fact, important and meaningful work for environmental historians. Recent environmental historians hew to the distinction between "old" and "modern" whaling and point rightly to the fact that old whalers and modern whalers hunted different types of whales, using different technologies, for the purpose of producing different commodities, and at vastly different scales of slaughter.[35] Even more recently, Bathsheba Demuth has modeled responsible environmental history in a careful comparative study of whaling practices in the Bering Strait: Indigenous whaling by Chukchi, Iñupiaq, and Yupik hunters, Soviet whaling, and US commercial whaling.[36]

Early twentieth-century Yankee whaling heritage makers created the sharp distinction between Yankee whaling and modern whaling for different reasons and through distinct historiographical strategies. Rather than draw on evidence or on the testimony of working whalemen, the Yankee whaling heritage makers leaned heavily on the comparison of Yankee whaling to sport hunting to differentiate it from modern whaling. The Yankee whaling heritage makers denigrated the technologies of modern shipping and whaling on the grounds that modern whaling was not "sporting" in the way that Yankee whaling in the nineteenth century had been. Yankee whalemen were cast quite literally, as the example of the whaleman statue shows, into the role of heroic sport hunters. Clifford Ashley even wrote sportsmanship into his specious origin story of US commercial whaling in *The Yankee Whaler*. He imagines that a right whale swam into the harbor of seventeenth-century colonial Nantucket and thus "aroused . . . the sporting instinct of the Nantucketer."[37] For Ashley, the innate "sporting instinct" of the white settler is the event that incites the commercial Yankee whale fishery, rather than the settlers' appropriation of Indigenous labor.

But despite such efforts by Ashley, Crapo, and others to cast whalemen as sport hunters, whalemen and the whaling industry fit awkwardly into the sport hunting analogy. To begin with, the diverse body of laborers who worked in the US whale fishery past and present did not at all resemble the wealthy white men who called themselves sport hunters. The comparison of whaling to sport hunting failed on environmental grounds, too. The industrial slaughter of whales within the context of a commercial whaling industry—even on the comparatively smaller scale of the nineteenth-century industry—violated nearly all the sportsman's rules about hunting for market and at a sustainable scale. In spite of these glaring inconsistencies, the creators of Yankee whaling heritage clung to the analogy and even wrapped whaling into the sport hunter's specious claims to conservation by implying that Yankee whaling had actually benefitted whales and protected oceans, in contrast with modern whaling, which, like mercenary market hunting, was practiced at an unsustainable scale. Yankee whaling preservationists contorted their ecological defense of whaling to an absurd degree. In turn, the sport hunter's claim that sportsmanship engenders conservation and stewardship fully unravels when it is applied to extractive industry at the scale of industrial whaling. The comparison between sport hunting and Yankee whaling was more than just an awkward fit: the analogy exposes the internal contradictions of the sport hunter's claims to conservation and reveals the violence at the heart of early twentieth-century sporting ideology.

An essay from 1932 encapsulates the whaling-as-sport-hunting analogy and the pseudo-ecological defense of Yankee whaling. The essay by British American writer William McFee serves as a foreword to the published collection book of whaling drawings, *Greasy Luck: A Whaling Sketchbook*, by the American artist Gordon Grant.[38] *Greasy Luck* offers a pictorial history of the nineteenth-century industry. Its drawings, each accompanied on the facing page by a long explanatory caption, portray scenes from the nineteenth-century US whaling industry. McFee's foreword differentiates the historical industry portrayed in the *Greasy Luck* drawings from the ongoing modern whaling industry: he celebrates the old industry for its sporting virtue and critiques the modern industry for its unsporting mechanistic efficiency. He does so in an extended comparison of whaling to various forms of pig slaughter: "The difference between this adventurous and romantic calling [Yankee whaling] and the modern whaleship is precisely the difference between pig-sticking as practised by army officers in India and the stock-yards of Chicago and Argentina."[39] The "pig-sticking" that McFee refers to here is the hunting sport taken up by British colonial officers in India and other sites under British colonial rule, in which hunters often on horseback would hunt wild boars and other feral pigs with long, pointed spears not unlike harpoons. The sport of pig-sticking was a grotesque crystallization of colonialism in its cruelty to pigs and colonial subjects alike; it was an especially offensive activity in places with large populations of Muslims or Jews who abstain from consuming pork on religious grounds. But to McFee, pig-sticking, like Yankee whaling, represents the apogee of sport. The close connection between McFee's ideal of sporting good fun and colonial military domination is telling and yields insight into the unspoken ideals that whaling represents for him: Yankee whaling of the nineteenth century was also a site for colonial domination, and whaling a type of sporting hunt that could be carried out by colonizers within the arena of occupied colonial territory.

For McFee, pig-sticking in India and Yankee whale hunting are good because they make the hunter/whaler direct agents of their own violence; modern whaling and industrial animal slaughter are bad because they alienate the killer from the act of slaughter. It is clear but it bears emphasizing that the only subject position that matters is that of the hunter; he does not name or acknowledge the life of the hunted animal. But McFee does acknowledge animal life on the scale of the species. McFee evokes the threat of whale extinction in order to critique modern whaling for the scale of its slaughter. "[Modern whaling] is so efficient that unless some legislative action is taken, whales will become extinct. In two years these vessels have obtained more oil and have killed more whales than the old American whalers took from the sea in half a century."[40] But McFee's sudden interest in animal welfare is specious; his overawing interest in the way whaling confers valor on the whaleman takes precedence through the force and vividness of his language:

> It [Yankee whaling] is sport because the hunters risked their lives when the harpoon left the boatsteerer's hands to plunge into the whale's carcase. They were in the most dire peril of a "stove" or a "chawed" boat until the animal's terrific struggles were ended by the thrust of a lance through his vitals. In modern whaling the operatives are in no more danger than the person who slits the jugular veins of the hogs suspended by their hind legs on a moving chain in a Chicago abattoir. I doubt exceedingly whether these modern whalers will ever have any songs or traditions.[41]

McFee's relish in the violence of nineteenth-century whaling is apparent in the vivid language—"plunge," "thrust"—he uses to describe the killing of the whale and in his adoption of the quaint whalemen's terms "stove" and "chawed." McFee's contempt for the comparative ease and riskless quality of slaughterhouse killing—and, by extension, of modern whaling—is clear. McFee's nod to the extinction threat of modern whaling appears, by contrast with the attention he pays to the valor of whaling, to be offered without conviction.

Praise for the risky "sport" of whaling is a belief that rests on the disregard of human life and well-being, too. McFee and other whaling historians praised and romanticized the risks inherent to Yankee whaling—of physical risk to life and limb and financial risk to spectators—without acknowledging the death, dismemberment, loss, and pain that are so often the end of risky ventures. Both physical and financial risk are subsumed under McFee's insouciant expression "sport," as if whaling were undertaken for the sake of physical and financial thrill-seeking rather than for the valuable commodity of oil that it produced. The language of sport in early twentieth-century whaling histories simultaneously acknowledges and diminishes the physical risk of whaling to whalemen. Whaling in all periods has been a deeply dangerous industry for the laborers who carried it out. In stark contrast to twentieth-century writers such as McFee, Herman Melville in *Moby-Dick* foregrounded the risk of death to whalemen. "Not a gallon [of whale oil] you burn but at least one drop of man's blood was spilled for it," Ishmael warns, and Melville begins his account of the whaling industry, not with idealistic vignettes of sporting whalemen, but with a somber picture of whalemen's widows gathering in a New Bedford chapel filled with cenotaphs.[42] Twentieth-century whaling histories are not somber when they consider the risk of death to the actual laborers; the reader finds heartfelt melancholy, grief, and loss in these histories only when they describe the metaphorical death of the industry and the apparent extinction of white Yankee whalemen. In the early twentieth-century histories of the industry, the romance of sporting risk displaced the grief felt by those—human and nonhuman—who experienced actual losses and actual deaths in the whaling industry. Extractivist nostalgia is available only to those who are protected by power from the losses of human and nonhuman life that extraction actually incurs.

The sporting history of Yankee whaling heritage helps us understand how we got the version of unsustainable energy production, unjust labor remuneration, and outright white supremacy that we in the twenty-first century are currently stuck in. Yankee whaling heritage follows the path of sport hunting in consolidating whiteness, wealth, and imperial violence with resource extraction. The violent ideology of sport lingers in contemporary business; "sport" has transmogrified into the "venture" of "venture capitalism," whose disciples retain the ghoulish celebration of "risk" in whaling history.[43] The language and visual iconography of "sport" haunts the representational history of petroleum, too. In 1901, the Standard Oil Company in coordination with civic leaders in Titusville, Pennsylvania, erected a monument to the oil industry that might be understood here as a sibling to Crapo's whaleman statue. The Titusville monument was meant to honor Edwin Drake, who has been credited with kicking off the Pennsylvania petroleum boom of the 1860s and 1870s, but it features a figure that might have been drawn from the sport hunting visual archive: a muscular man—an allegorical figure called The Driller—who crouches at the site of an oil well, frozen holding a hammer high in his right hand over the stake at the drill site. Like the harpooner, the driller is pictured in the instant before he commits the direct act of violent extraction; both figures are heroic because of their capacity to inflict extractive violence.[44]

But even in the bonanza of extractivist nostalgia that is McFee's essay, the story of energy is more complicated. Unlike most of his contemporaries, McFee acknowledges the problem of environmental harm when he invokes extinction. And although his extinction anxiety is clearly specious, he appears prescient in the way he connects attributes extinction with anonymous mechanization of animal slaughter in modern, fossil-fueled whaling ships. Whaling is and always was part of an extinction-producing economy, and McFee apprehends at least that part of the industry's danger.[45] McFee is blinkered by his nostalgia for situations in which white men like him feel licensed to commit violence, but it impossible to ignore his genuine if imperious horror at the prospect of slow violence and its origins in extractive capitalism: "Here we behold the modern commercial and mechanical genius at work. It is so efficient that unless some legislative action is taken, whales will become extinct."[46] McFee's unvarnished delight in imperialism and direct violence coexists in this essay with his apprehension of the slow violence of extinction. But in encompassing these seeming contradictions, McFee's essay reveals the contingency of the very concepts of energy, extraction, and extinction.

DOWN TO THE SEA IN SHIPS

Extractivist nostalgia for the New Bedford whaling industry found its fullest and most complex expression in the silent film *Down to the Sea in Ships* (1922). It was finally through a fictional narrative form, a motion picture, that the self-appointed custodians of New Bedford whaling history were able to present their vision of nineteenth-century Yankee whaling as a field of sportsmanship and a bastion of white supremacy. The film stands alongside local histories and civic monuments as a product of historic preservation efforts in New Bedford because the film was financed and produced by a group of wealthy investors in New Bedford; *Down to the Sea in Ships*, too, was a civic monument. In October 1921, a group of New Bedford citizens established the Whaling Film Corporation, whose purposes were stated as follows: "The production and exhibition, with appropriate accessories, of a motion picture depicting by means of a story the whaling industry as formerly conducted in New Bedford (Mass.)."[47] By late 1921 the Whaling Film Corporation engaged as director Elmer Clifton, a protégé of the director D. W. Griffith. Clifton worked with Griffith on Griffith's white supremacist film *The Birth of a Nation* (1915), which portrayed Black characters using grotesquely racist stereotypes and romanticized the rise of the Ku Klux Klan as a force that restored order in the Reconstruction South. *The Birth of a Nation* was widely decried by Black critics in its time and ours. But the whaling film made by Griffith's protégé has thus far not received the same attention. In *Down to the Sea in Ships*, Clifton deployed the inventive cinematic techniques that Griffith had pioneered, as well as an anti-Indigenous, anti-Black, anti-immigrant, white supremacist view of American history that resembled that in *The Birth of a Nation*. The picture of Yankee whaling as a bastion of white supremacy appeared in the film as seamless and frictionless because it was utterly fictional. *Down to the Sea in Ships* was released in 1922 and became popular nationwide, although it is best known in film criticism today for being the first film in which the film star Clara Bow had a significant role.

The film's plot is a melodramatic labyrinth of thwarted love, shipboard mutiny, racial passing, and the struggle of an old generation to keep its racial heritage white. The film is set around 1850 at the height of productivity and prosperity in the New Bedford whaling industry. The narrative centers on the patriarch of a venerable Quaker whaling family, William W. Morgan, whose pride in his white colonial ancestry is his defining personality trait. William Morgan has a big problem, because Morgan has no son. Morgan once had a son, who died at sea, and now he has only an unmarried daughter named Patience. Without an heir to the Morgan whaling fortune and family name, Morgan feels exiled from both past and future; he is a disappointment to his American ancestors and a liability to the nation. The film stages Morgan's anxiety by showing him alone in his attic, where he digs around in an old trunk to find an American flag. The risk that Patience will marry badly and fail to perpetuate her father's racial heritage and belief system is the central conflict of the narrative. Two different hazards threaten the futurity of Morgan's whiteness, each taking the form of a different suitor to Patience. The first unsuitable suitor is Thomas Allan Dexter, an industrialist whose cotton mills stand for the new steam-powered industrial order that threatens New Bedford's older and more virtuous extractive natural resource industry. Allan asks for Patience's hand in marriage, but Morgan rejects the union in a speech given on one of the film's title cards: "Patience is a whaleman's daughter. Unless thee has thrown a harpoon into a whale, take thy story of love elsewhere. It can never be—never!" Allan's whiteness is not enough until it is tested and proved in the crucible of natural resource extraction. Allan, of course, ends up throwing harpoons on a whaling voyage.

The second threat to the successful perpetuation of Morgan's family and his values comes in the form of another suitor for Patience: Samuel Siggs, who represents the nonnative immigrant threat to racial purity. This suitor appears perfect to Morgan: Siggs appears to be a successful agent in Morgan's counting house, a former whaleman, and a dutiful Quaker. But Siggs is a Chinese imposter disguising himself as a "native" New Englander in order to infiltrate Morgan's business and take over the family fortune by marrying Patience. Siggs first appears on film as an orientalist stereotype, wearing an embroidered silk dressing gown and holding a lacquered fan. Later in the film, he dresses in Quaker disguise. A title describes Siggs's concealed Chinese identity as a "sinister yellow strain hidden by sheep's clothing." Eventually, Siggs is unmasked as a Chinese immigrant and banished from the community, while Allan completes a successful whaling voyage. Allan marries Patience, Patience gives birth to a son, and the purity of Morgan's family heritage is vouchsafed for the future through his son-in-law's whiteness, proved by his whaling prowess. The film closes on an image of Morgan's baby grandson resting in the cradle handed down by Morgan's ancestors.

The film is set in the mid-nineteenth century, but its anxieties are all of the twentieth century. Morgan is a proxy, *not* for the whaling scions of the nineteenth century, but for the civic boosters spreading white supremacy through the revival of whaling history in the twentieth. The cultural texts from the Yankee whaling revival practically vibrate with anxiety about the status of white masculinity and the future of whiteness. The film's patriarch, Morgan, on his knees clutching the American flag in his attic, resembles Clifford Ashley's painting from 1911 of an old whaleman kneeling in front of his sea chest.[48] Morgan's anguished quest to find a white Quaker whaleman to marry his daughter, Patience, represents his reckoning with the limits of his own masculine virility and with white supremacist fears about "race suicide" or "white genocide." Images of weak masculinity and wasted fertility suffuse whaling narratives about the end of the industry. Ashley's narrative history *The Yankee Whaler* evokes lost masculine virility in the image of abandoned oil casks on the New Bedford waterfront: "In the years that were to follow [the petroleum boom], the seaweed blew away unheeded, the hoops rusted, the staves dried out, and his [one impoverished shipowner's] oil trickled and seeped down into the earth of the wharf unnoticed."[49]

The film was the cinematic equivalent of the whaleman statue: a white fabrication of history that filmmakers invented through strenuous effort when actual historical facts would not serve white supremacy. But as with the creation of the whaleman statue, the process of bringing this white supremacist version of whaling history to film was an enormous job whose sole means of success was also its chief obstacle: the actual, ongoing New Bedford whaling industry, which was operated by a racially diverse body

Clifford Ashley's painting of an old sailor in an attic evinces the same kind of extractivist nostalgia as the silent film *Down to the Sea in Ships*. Clifford Ashley, *The Sea Chest*, 1911. Courtesy of the New Bedford Whaling Museum.

of laborers, officers, and owners. In order to gain access to whaling ships, whaling artifacts, and, above all, the embodied knowledge of whaling labor, the white supremacist filmmakers had to work with the people that their film disparaged and erased. In the spring of 1922, the members of the Whaling Film Corporation, who called themselves its "captains," hired a fishing boat and a crew of New Bedford whalemen to take the actors and film crew on a whaling expedition to the Caribbean in order to film the "authentic" whaling scenes that promoters saw as the climax of the film's action: the scene where Allan Dexter finally earns the right to Patience's hand and, by extension, proves his whiteness. Film promoters touted the film's realism and its authority as a true document of Yankee whaling as its main selling point. The film distributor's promotional materials offered up a reaction formation: "The most astounding piece of realism ever photographed—a motion picture made in the broad reaches of mid-Atlantic, with a 90-ton sperm whale as the principal actor, and the hand of chance directing as fierce a battle between man and animal as the long history of whaling has ever known."[50] A look at the film's making reveals the lengths to which the filmmakers went to whitewash the history of the whaling industry.

For example, the whaling voyage that the film crew undertook was nothing whatsoever like the scenes pictured in the films. The work of whaling in the film was made possible by laborers of color, whom the filmmakers literally edited out of their shots at every opportunity, and by all of the modern technologies that Yankee whaling heritage makers eschewed as unsportsmanlike. The expedition was a motley assemblage of New Bedford whalemen and a Hollywood cast and film crew crammed together on a gasoline-powered schooner called the *Gaspe*, together with a small nineteenth-century wooden whaleboat, harpoons, and up-to-date camera equipment. The promoters of the movie touted the transformation of the star, Raymond McKee, from a Hollywood dandy who showed up to New Bedford "attired in white flannels and carrying a golf bag under [his] arm" into a seasoned whaleman.[51] McKee and the film distribution company fashioned his persona into that of a real whaleman, like the character he portrayed, who earned his star turn through the initiation of extractive violence. There is even in the collection of the Mystic Seaport Museum a scrimshawed sperm whale's tooth attributed to McKee, featuring a portrait in tooth and ink of Clara Bow, Clifton, and several other members of the film's cast and crew. Apparently enamored of whaling lore, McKee learned to scrimshaw and covered his whale's tooth with images of the movie's setting and actors. At the bottom corner of the tooth is the image of A. G. Penrod, the movie's cameraman, and a cartoonish movie camera. Behind him with a bullhorn and sunglasses is an image of Elmer Clifton, and on the opposite side of the tooth, portraits of characters in costumes.[52]

During the making of the film, whales were indeed killed and expeditions were undertaken, but the voyage of the *Gaspe* was in actuality, according to the crew, "a great sporting adventure" and a project doused in the petroleum-fueled technology of the twentieth century.[53] But so great were the intellectual and political pressures on the film crew to render nineteenth-century whaling scenes in the form of their imagination that they brought all of the tools of petromodernity to bear on the illusion of realism. One of the film crew, Gordon S. Blair, kept a cheeky diary in the form of a bastardized ship's log and chronicled the sleight-of-hand tricks that the film crew used to make the whaling scenes appear real. In order to stabilize the camera for scenes shot at sea, the film jury-rigged a gyroscopic stabilizing tripod built out of "the universal joints from automobiles and a few hundred pounds of cement—the only available materials in Santo Domingo."[54] Those building blocks of petromodern automobility, car parts and cement, underscore the degree to which *Down to the Sea in Ships* is what Stephanie LeMenager has called a "petroleum medium" and not the paragon of nineteenth-century realism or whale oil culture that the

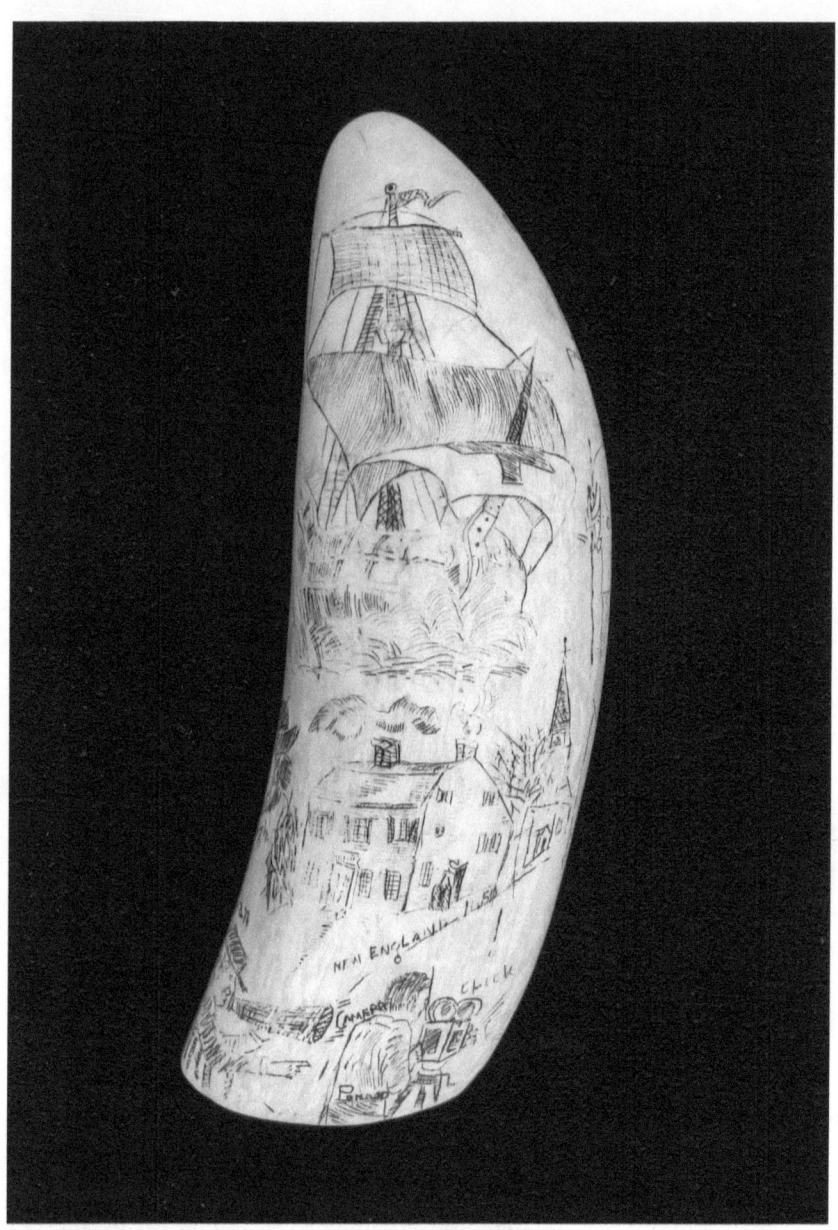

This scrimshawed whale's tooth was carved by the star of *Down to the Sea in Ships*, Raymond McKee, in 1922. Scrimshaw was one of many projects that the film's cast and crew undertook to authenticate themselves as "real" whalemen.
© Courtesy Mystic Seaport Museum, Collections Research Center.

film claimed to be.⁵⁵ The climactic scene where McKee's character finally harpoons his whale was a complete mock-up; McKee stood in a boat in the shallow waters of a Haitian beach and threw harpoons into a dead whale propped up in sand. Blair describes the setup in his log entry from March 26, 1922 (a day whose weather Blair notes as "Warm as Hell"): "Went ashore in the morning and started to take pictures of the whale which was mounted up. My job was to make waves with a paddle. The motor boat was backed up to the nose of the whale to make sea effect. A hose was used to make a spout. Took pictures of McKee harpooning whale. Ate lunch on beach from [blank space]. Bought some plantain on credit."⁵⁶ It is not for mere sport here that I puncture the film's claim to realism but to show the great distance between the imagined white Yankee whaling world of sport and honor that the filmmakers wished to evoke and the world they actively inhabited. The whaling industry is and always was a technologically heterogenous space and one operated by people of strikingly diverse identities whose motives were survival on the part of the laborers and profit on the part of the investors.

The film's historical inaccuracies and flamboyant fabrications were perhaps only an outgrowth of Clifton's own self-deception. Elmer Clifton seems to have undertaken a huge amount of research into New Bedford whaling history. He consulted literary critic Raymond Weaver's Melville biography *Herman Melville, Mariner and Mystic* (1921) and mid-nineteenth-century maritime narratives such as J. Ross Browne's *Etchings of a Whaling Cruise* (1846) and the even more obscure narrative *Salt Water Bubbles, or Life on the Wave* (1854), by Hawser Martingale.⁵⁷ Clifton worked closely with the Old Dartmouth Historical Society, and Clifton corresponded with local living historians including Albert Cook Church and Clifford Ashley. Ashley reviewed Clifton's script of the film and sharply criticized its anachronistic history of the New Bedford whaling industry:

> You should not, early in your story, make too much of the idea of the modern age supplanting the whaling. The fact is that whaling was still in its ascendency [in the mid-nineteenth century, when the film is set], the peak had not been reached. Neither of your two young men symbolize the age to which they belong. One is a commonplace character, the other is mechanical and fears the water. Since you are writing the motion picture whaling epic, do you not think this is a mistake? Do you not think that the worth while youth of that period would have been wholly dominated by the life of adventure at their very doors? At any rate, don't you think it would be more dramatic to have them so stirred? Don't you think what you have seen of whaling

as it is at present carried on, has unconsciously influenced you? That is, that you fail to see whaling as it was then?[58]

The last two questions signal that Ashley's main critique is that the film's sensibility is born of twentieth-century anxieties rather than nineteenth-century realities—that Clifton has been "unconsciously influenced" by "what [he] has seen of whaling as it is at present carried on" (that is, in its decline, carried out mainly by foreign-born people of color) than what Ashley and others have told him of what it was in the bygone days of Yankee whaling. Clifton did not take Ashley's suggestion to revise Allan's characterization to better fit nineteenth-century economic realities. The film's central conflict is still driven by white supremacy of the twentieth century, which in narrative form dramatized the clash of old and new, foreign and native-born, white and other. And given the latent white supremacy in Ashley's own work, it seems that his critique owes in part to the narcissism of small differences.

ENDLINGS: THE *Wanderer* AND THE *Morgan*

The white supremacist anxiety about race extinction motivated the creators of New Bedford civic monuments dedicated to the memory of Yankee whaling: Crapo's whaleman statue of 1913 and the *Down to the Sea in the Ships* film of 1922. Those anxieties came to be attached, too, to the infrastructures and artifacts of whaling that still littered the New Bedford landscape. I wrote in chapter 2 about how late nineteenth-century tourists reflected on the picturesque dilapidation of New Bedford whaling ships and warehouses while transiting quickly to the modern steamships that would carry them to the more fashionable seaside resorts of Nantucket, Martha's Vineyard, or nearby Dartmouth. Obsolescing whaling infrastructure indexed New Bedford's decay and decline. But in the 1920s, the sight of those obsolete vessels and infrastructures evoked among New Bedford's wealthy preservationists a defensive pride in the city's once-glorious Yankee whaling past—one that ought to be celebrated and preserved for posterity. The same might be said for the frantic collection, curation, preservation, and exhibition of all the artifacts that filled the Old Dartmouth Historical Society and the New Bedford Whaling Museum in the first decades of the twentieth century. But by far the most charismatic and emblematic artifacts of the whaling industry were its wooden sailing ships.

From the point of view of heritage makers, the last chapter in the story of Yankee whaling begins with the end of two whalers: the *Wanderer*, which perished in a stormy wreck, and the *Charles W. Morgan*, which survived in the strange hereafter of an enshrined monument. Both ships were called the "last ship" of the Yankee whaling fleet, the last of their kind.[59]

The "lasts" of the Yankee whaling industry—the last Yankee whaling ships, the last Yankee whalemen—took on the valence of twentieth-century race science and extinction anxiety: a pseudo-ecological language casting the decline of Yankee whaling as a kind of ecological tragedy. William Crapo, in making his case for the whaleman statue in downtown New Bedford in 1913, evoked ecological collapse in describing the decline of Yankee whaling: he wrote that "the old stock is seriously diminishing in numbers," as if he was talking about pandas or cod. And the last ships—the *Morgan* and the *Wanderer*, were treated as what we in the twenty-first century would call the last of their species, or "endlings," to use a term that did not yet exist.[60] Once identified as the last of their kind, the whaling ships were anthropomorphized, mourned, and otherwise fussed over as individuals freighted with the weight of their species: they were charismatic endlings like George and Martha, the last known passenger pigeons, who died at the Cincinnati Zoo in 1910 and 1914, or Lonesome George, the Pinta Island tortoise from the Galápagos, who died in 2012.

In August 1924, the endling ship *Wanderer* prepared to set sail from New Bedford on a whaling cruise to the South Atlantic. For weeks leading up to the *Wanderer*'s launch, New Bedford residents and summer visitors came to see the whaler at the pier; most believed that the voyage would be the last whaling voyage to depart from New Bedford. The local newspaper reported the ship's impending launch as the end of an era:

> Color is making way for Power. Romance is being superseded by Efficiency. Sail has been pulled down and the steam-throttle has been opened up. The whaler of the olden day has been displaced by the more prosaic but effective steam-whaler, and the bark *Wanderer* is about to set out upon her last voyage.
>
> That is to say, from all appearances it seems that it may be the last voyage of the staunch old craft, and with her disappearance all of the once glorious fleet of New Bedford whalers will have faded into ghost ships or been reduced to sad, silent hulks, neglected except by the casual visitor.[61]

The *Wanderer* sailed out of New Bedford on Monday, August 25, 1924. Late at night on Tuesday, August 26, in a storm, the *Wanderer* crashed off the west end of Cuttyhunk, an island in Buzzard's Bay, just off New Bedford. From shore on the morning of August 27, the bark appeared intact, its mast standing high. But the hull was irreparably damaged, and when the captain came ashore, he declared that "her days of usefulness were ended."[62] The story led the first page of the *Boston Daily Globe*, and the eulogies—for the bark and for the US whaling industry at large—poured in during the days after. The ship's demise on the rocks of Cuttyhunk put a grim but gratifying punctuation mark on the history of New Bedford whaling. In the days after the wreck, locals and summer visitors lined up on the Cuttyhunk beaches to watch the salvagers bring soggy tobacco, oilskins, boots, and clothing from the wrecked ship back to shore. The *Wanderer*'s voyage had already been preordained its last, and its wreck gave the ship and the industry a fittingly operatic ending.

But like so many other "lasts," the *Wanderer* was not exactly the last whaling ship. While the *Wanderer* fell to splinters offshore, the whaler *Charles W. Morgan* stood empty and disused on a dock in Fairhaven, "dying by inches." In October 1924, just a few months after the *Wanderer*'s wreck, a local millionaire announced that he would finance the restoration of the whaling bark. The *Morgan*'s rescuer was "Colonel" Edward Howland Robinson Green, who planned to install the whaler on his estate at Round Hill, Massachusetts, as a public museum and a memorial to the whaling industry.[63] Green credited the wreck of the *Wanderer* with the inspiration for his memorial:

> When on the day following the recent storm I looked over the bay with a pair of powerful glasses which I had frequently tested by focusing on the bark *Wanderer* after she went ashore at Cuttyhunk, I could not pick her up. She was gone. Then and there at that moment it came over me that the *Charles W. Morgan*, rusting at her moorings in Fairhaven, was the very last relic of a glorious past.
>
> When she is gone there will be nothing to mark the industry. I immediately left the shore and came up to the house and dictated a letter to Mr. Neyland [the legal owner of the ship], informing him that if the city of New Bedford did not see fit to take over the *Morgan* I would do so. Then I started in operation a research which I hope will bring me historic facts which may be used in connection with the restoration of the vessel.[64]

Green provided for the restoration of the *Morgan* by forming an association with a fittingly funerary name: "Whaling Enshrined, Inc."[65]

Though Green is not as famous a figure as Carnegie or Rockefeller, like those families, Green and his fortune are emblematic of the links among extractive energy, Wall Street, and cultural patrimony in the late nineteenth and early twentieth centuries. His wealth derived from his family's long-term investment in extractive energy, in both the nineteenth-century New England whaling industry and twentieth-century fossil fuels. Green had wealthy ancestors in his grandparents' generation in the New Bedford whaling industry, and in fact, some of his family held partial ownership in the *Morgan* during its whaling years. Green's mother was Hetty Green, a powerful investor popularly called the Witch of Wall Street who had spun her own whaling inheritance into a fortune valued at $200 million at her death.[66] Her son Edward established his career in Texas, where with his inherited wealth he bought and operated the Texas Midland Railroad. Green invested in a wide range of interests in the Texas boom economy at the turn of the twentieth century—oil, roads, and real estate—and he was said to have bought in 1899 the first gasoline-powered automobile in Texas.[67] Green spent his money, in part, in the flashy mansion he built on property at Round Hill, in a tony district near New Bedford, in 1921. He developed the surrounding farm and property to suit his scientific hobbies: radio, airplane, and dirigible flight. Green donated funding and space on his property to scientific researchers: to MIT, which built fog and radio research facilities, and to the US Department of Agriculture and the Agriculture and Mechanical College of Texas, which carried out research on boll-weevil-resistant cotton plants.[68] On his estate at Round Hill, Colonel Green assembled two radio broadcasting stations, "machinery for smashing the atom," a "fog dispeller," "radio direction-finding equipment," and a dirigible with its own hangar.[69]

With its antique wooden technology, the whaleship *Charles W. Morgan* was an odd addition to the technological marvels on Green's estate, and the vessel was explicitly differentiated from the technological menagerie at Round Hill by its status as a dead thing. During the two-year restoration project, the ship was towed from its berth in Fairhaven to a protected dock at Round Hill and pressed into a bed of cement, an installation that newspapers described as an entombment. The *Boston Daily Globe* described the ship as "permanently encased in a resting place of concrete," and several newspapers followed the language of Whaling Enshrined to declare the site a "shrine" to whaling history.[70] In fact, Whaling Enshrined was a double monument of sorts: the mast of the *Wanderer*, salvaged from the wreck at Cuttyhunk, was resurrected on shore and installed near the *Morgan*.

The *Morgan* was dedicated at a festive ceremony on July 22, 1926, and opened to the public as a free museum. Green opened the dedication ceremony with a speech in which he "welcome[d] home the good old ship *Charles W. Morgan*" and eulogized the whaling industry as an energy industry, "through which all of our kin folks worked so hard to bring New Bedford and the surrounding country prominently before the world and at the same time furnishing oil for the world and in this way helping to enlighten it." The Whaling Enshrined museum site was a success, bringing in an estimated two million visitors in the ten years following its dedication.[71]

Some of the *Morgan*'s visitors themselves became part of the tourist attraction. The ship was reported to have served as a meeting place for retired whalemen, whom visitors sought out alongside the ship's harpoons and try-pots as living relics of the whaling days.[72] And for visitors and veteran whalemen alike, the exhibition was designed with an olfactory exhibition meant to raise old memories. The dock alongside the *Morgan* was piled with casks of whale oil, put there expressly to infuse the surrounding air with its unique scent. "Col. Green has stored casks in seaweed by the side of his ship that they may bring back old memories of departed days to the old citizen who gets a whiff of oil and seaweed. He proposes to keep his casks saturated in whale oil, restoring the distinctive smell of New Bedford a century ago."[73] The *Morgan* with its garland of oily casks would be a postindustrial madeleine that transported its visitors into the obsolete past.[74] In his book *The Yankee Whaler*, Clifford Ashley testified to the power of that scent: "As for sperm-oil, I cannot smell it to-day without an attack of nostalgia; the faintest whiff and I see again the old New Bedford wharves, black with oil-soaked earth, rough-binned with seaweed-covered casks; and fringed with long rows of dismantled square-riggers, their jibbooms housed and yards cock-billed. Give me the smells of a spar-yard, a rigging-loft, or a New Bedford wharf and all the refinements of the perfumer's art hold not a single charm."[75] The ship flying with festive flags, staffed by old whalemen swapping stories, and redolent of the scent of whale oil was an expression of extractivist nostalgia for the full sensorium.

Whaling Enshrined and the *Morgan* offered, in miniature, something like the odd historical assemblage of historical buildings and artifacts that Henry Ford built at precisely the same moment at Greenfield Village in Michigan. Historian Steven Conn locates the irony of efforts like Green's and Ford's: nostalgic tributes to the past that, in Ford's case, were "collected by the man whose automobiles and factories, perhaps more than anything else, caused that past to vanish."[76] The whaling world that Green paid homage to with the *Morgan* and Whaling Enshrined was precisely the world that his and his mother's accumulation of capital through Wall Street investment and petromodern speculation had made obsolete. Historical preservation efforts such as Greenfield Village and Whaling Enshrined are born directly of the extractivist nostalgia that mourns what it destroys.

A ship is not a narrative, and the *Morgan* does not tell of the same explicit white supremacist ideology as the film *Down to the Sea in Ships*.

But in its instantiation as "Whaling Enshrined," its concrete entombment at Round Hill from 1926 to 1941, the ship became part of a larger local narrative of industrial change as one of extinction, decline, and death. The white supremacist anxiety of racial extinction was loaded like freight onto the *Charles W. Morgan* and, to some extent, onto other nineteenth-century wooden sailing ships like it that to this day in the United States are persistently associated with whiteness.

CODA: TRUMP DIGS COAL

The early twentieth-century whaling history revivalists told nostalgic stories about extractive industries that were also fictional parables of white grievance. Theirs were some of the narratives about energy transition that keep us stuck in fossil modernity and in the systems of injustice that fossil modernity produces. As Mark Simpson and Imre Szeman warn: "Transition's fiction thereby makes sure that we remain stuck in a present that withholds, in the active language of its idiom, any capacity to create a genuinely different energy future."[77] This narrative form of extractivist nostalgia persists to the present.

The "Make America Great Again" politics of Donald J. Trump are, as many have observed, deeply nostalgic for an imagined past of uncontested white patriarchal rule: lives built around "cars, suburbs, and the nuclear family" and made materially possible by an ongoing supply of cheap fossil fuels.[78] Cara Daggett, Dominic Boyer, Cymene Howe, and Timothy Mitchell put the centrality of fossil fuels to Trump's vision by calling it "petro-nostalgia."[79] But however apt, the terms "petro-nostalgia" and "petro-masculinity" do not quite capture the specific role of the obsolescing energy industry in Trump's nostalgic ideology. Trump's fixation on coal during his presidential campaign of 2016 follows the rhetorical path forged by the Yankee whaling heritage makers a century earlier: they coopted extinction grief not for the victims of extractive industry but for their perpetrators.

During his presidential campaign, Trump declared an end to the government's "war on clean, beautiful coal," and at several moments during Trump's presidency, the administration announced new coal-friendly policies that lower pollution standards at power plants. But, of course, coal is on the way out. The coal industry employs fewer people than the Arby's restaurant corporation or Walt Disney World.[80] The solar energy industry employs five times more people than the coal industry.[81] It is clear to environmental activists and energy industry analysts alike that coal extraction is ending. The comprehensive report issued by the International Panel on Climate Change in 2018 made it clearer than ever that eliminating coal-fueled power plants is an essential step in mitigating catastrophic global warming. Still, Trump digs coal. Trump's views on coal are profoundly nostalgic: Trump wants to bring the coal industry back and make it, like the rest of America, great *again*.

Like the brawny harpooners of the Yankee whaling industry, working-class coal miners are ostensibly the heroes of the nostalgic story that Trump tells through specious personal memories: "The people that like me best are those people, the workers. They're the people I understand the best. Those are the people I grew up with. Those are the people I worked on construction sites with."[82] In his role as president, Trump projects his nostalgia for "growing up" with laborers onto the big screen of national history in order to fantasize an American past in which millionaire (white) financiers and noble (white) laborers worked side by side in the interest of (white) American greatness.

For the New Bedford historic preservationists, extractivist nostalgia took the form of funeral-like commemorations; Trump, meanwhile, is not content with a funeral and has said that he wants to revive coal. But Trump did not actually effect any new policy to bring coal back or protect its jobs; nearly 10 percent of coal jobs were eliminated during the Trump presidency between 2016 and 2020.[83] Yankee whaling nostalgia and coal nostalgia seemingly make heroes of working-class laborers, but the empathy is shallow at best. The call to revive coal was symbolic, and according to coal historian Peter Shulman, "Trump's pledges to coal miners were rhetorical appeals to hard-working, blue-collar Americans."[84] "Hard-working" and "blue-collar" in Trumpian discourse stand for "white." Like historic preservation in New Bedford, Trump's coal nostalgia operates only in the realm of culture and rhetoric, where its main purpose is to shore up the myths of whiteness by telling the story of a tragic ending.

In the case of both whaling and coal, the twinned efflorescence of extractivist nostalgia and white supremacy at the site of declining resource regimes is not genuine grief over extinction but a phenomenon that psychologists call an "extinction burst."[85] When you are trying to wean yourself off a destructive behavior (such as, say, smoking, binge drinking, or extractive capitalism), you might experience an intense relapse of the behavior just before eliminating that bad habit for good. It is worth attending to those who view energy obsolescence as a type of extinction event: the grief that communities experience when industries depart is real. But conflating these changes with the scientific and ecological concept of "extinction" opens up an ideological pathway for white supremacy, and it explains in part the surprising eruption of the misplaced term "extinction" in Yankee whaling heritage.

As we have seen, efforts to commemorate the bygone Yankee whaling industry in New Bedford were in fact public exhortations of white supremacy and nativism. White elegies for whaling based their emotional and ethical force through racializing the story of economic change and coopting such ecological terms as "extinction" to make demographic change seem catastrophic. I again follow Rosaldo in my desire to explore and expose the logics of extractivist nostalgia: to "dismantle" rather than simply "demystify" the ideology of white supremacy.[86] By recognizing how "extinction" has in the past been freighted with white supremacy, we can disentangle the two terms and accord the very real threats of extinction and other forms of climate-change induced harm the attention and concern they warrant. Existential threats are at hand; other forms of survival and grief are possible.

5

Nostalgia for the Wooden World
Energy, the Melville Revival, and Rockwell Kent's *Moby-Dick*

In 1851, when *Moby-Dick* was first published, the whaling industry reached peak whale oil and fossil modernity loomed. In the early twentieth century—in the era of oil trusts and trustbusters, Ford factories, Model Ts, flying aces, gas-powered warships, and home electricity—*Moby-Dick* became a different book.

Moby-Dick became, for the first time in its existence, a wildly popular book. The novel that had been largely dismissed in Melville's own time was reappraised, named a masterpiece, and widely republished and read in the phenomenon of critical reassessment that critics have called the Melville Revival. The story of the Melville Revival has been told many times, and the landmarks of its so-called first wave are conventionally marked as follows: 1919, with public recognition of the centenary of Melville's birth; 1917–21, with the publication of the four volumes of the first *Cambridge History of American Literature* by professors at Columbia University; 1921, with the publication of the first critical Melville biography, Raymond Weaver's *Herman Melville: Mariner and Mystic*; 1929, with Lewis Mumford's biography *Herman Melville*; and 1941, with the publication of F. O. Matthiessen's *American Renaissance: Art and Expression in the Age of Emerson and Whitman*.[1]

In the century since that first wave, the Melville Revival has itself become the object of history and criticism. What interested most critics of the Melville Revival was Melville's Americanness. What has interested later critics who study the Melville Revival was the relationship between the Melville Revival and the institutions of US literary studies that emerged around the same time. Melville and the Melville Revival have served as bellwethers for the politics of the United States and the changing status of US literary studies in US political and cultural life. What is Melville's vision of America? What should Melville's place be in the canons and institutions of US literature? And what does the changing critical reputation of Melville and *Moby-Dick* help us understand about the politics of US literary studies?² Over the years, Melville scholars have followed and shaped the complex political currents within US English departments.³ As Eric Aronoff nicely puts it: "The study of Melville studies, then, could be said to be the study of American literature as a discipline."⁴

I would like to tell a different story about the Melville Revival, one in which *Moby-Dick* and the Melville Revival mediate energy transition and the tectonic changes wrought by fossil-fueled energy in the early twentieth century. The Melville Revival, in my energy-centered view, reached its apogee in 1930, with the publication of a dazzling new edition of *Moby-Dick* illustrated by the artist Rockwell Kent. Kent's *Moby-Dick* ought to stand alongside Weaver's and Mumford's biographies as an important landmark in the Melville Revival for three reasons: (1) because Kent's wildly popular illustrated edition brought many new readers to the novel; (2) because Kent's signature woodcut style became the signature visual style that denoted "classic" American literature; and (3) because Kent's *Moby-Dick* ought to be central to energy-oriented and, more broadly, environmental readings of the novel.⁵ I read Kent's *Moby-Dick*—in the form and content of his illustrations and in the method of their production—as a coherent if complex meditation on energy, work, and technological change, an important work of Melville interpretation as well as a landmark in Kent's own lifelong preoccupation with these topics. Kent's *Moby-Dick*, through its form and design as well as in its relationship to Melville's text, also creates a visual language for the cultural process of obsolescence that I have described throughout this book. Kent's *Moby-Dick* renders obsolescence across several scales: the obsolescence of old cultures of energy and print and of the wooden world of ships and the sea.

At a moment when other artists grappled with fossil modernity by representing cars, airplanes, oil wells, and factories, Rockwell Kent in the *Moby-Dick* illustrations made sense of fossil modernity by exploring the labor, technologies, and infrastructures that fossil fuels were making obsolete. Cara Daggett has argued that the mass extraction of fossil fuels in the Anglo-American world in the mid- and late nineteenth century gave rise to a then-new concept of "energy" inextricably yoked together with "work"; the "energy-work nexus" then gave rise to new forms of labor and governance and to the intensification of fossil fuel extraction and consumption.⁶ In the *Moby-Dick* illustrations and elsewhere in his oeuvre, Rockwell Kent works at the energy-work nexus as if it were a knot, trying to loosen the bindings of work, worth, and value from fossil modernity. Kent's *Moby-Dick* reveals just how tightly the sutures of fossil-fueled energy and work are bound in fossil modernity.

Moby-Dick, as I have argued in chapter 1, has always been about energy and extractive capitalism. Melville's novel described and critiqued the features of extractive capitalism that had already emerged in US commercial whaling and that would characterize the fossil fuel energy regime to come. *Moby-Dick* imagined many limits and ends to the whaling industry he chronicled: the extinction of whales (and humans), resource exhaustion, and obsolescence. The novel's restless cycling through different end scenarios for the whaling industry was its form of critique. By proliferating these endings, the novel suggests that capitalism's mandate for endless growth is impossible, especially in extractive economies that deplete the future world.

Those end scenarios resonated differently in the early twentieth century, as the form of the industry that Melville chronicled—the commercial whale oil industry, with its wooden sailing ships—actually ended. As I chronicled in chapter 4, the US whaling industry's end was publicly mourned and commemorated, especially in New England port cities such as New Bedford that had been the industry's centers.[7] Those public commemorations of whaling were wrapped into a narrative of energy transition, where the old, organic energy regime of whale oil, wind, and human muscle was contrasted with the mechanization of coal, oil, steam, and iron. The story of energy transition in whaling commemorations served conservative and reactionary ideologies: the fiction of an all-white whaling industry called Yankee whaling was invented at the moment of that industry's supposed death in order to stoke the fires of white supremacy, masculinism, nativism, and extractive boosterism. Indeed, a form of the industry was ending, but its twentieth-century commemorators conveniently ignored the industry's multiracial history and its volatile booms and busts in order to link energy transition with white aggrievement. *Moby-Dick* appears different in this new context, available to readers who can ignore the novel's critique and approach the ends of the industry with the extractivist nostalgia that drove some of those public commemorations.

My energy-centered account of the Melville Revival does not fully diverge from the historiographic narrative I sketched above. By putting the Melville Revival into the context of a larger national whaling history revival, I do stress the conservative vein of thought in these revival efforts: the ways that the history of a bygone resource regime like whaling, when viewed through the lens of nostalgia, becomes available to white supremacist and nativist views of US history. And indeed, the Rockwell Kent

Moby-Dick, especially when considered with the rest of Kent's oeuvre, in which one sees many glorifications of settler colonialism, can be understood as a work of extractivist nostalgia.

But the story that emerges through an energy and Kent-centered account of the Melville Revival is much more complex. To begin with, Kent's *Moby-Dick* reveals that the politics of Melville, *Moby-Dick*, and the Melville Revival in any period cannot easily be claimed for the right or the left. Kent was an avowed leftist, although I argue that his *Moby-Dick* in some ways aligns with the reactionary ideologies he professed to hate.[8] Through the form, content, and method of his illustrations, Rockwell Kent's *Moby-Dick* evokes intense nostalgia for strenuous labor, which he locates in two idealized sites of representation: in the pre-fossil-fuel whaling history that Melville's *Moby-Dick* represents and in remote regional coastal villages, where old forms of maritime labor appear to persist outside the ken of fossil modernity. In the illustrations for *Moby-Dick* and throughout his oeuvre of painting, printmaking, and writing, Kent interrogates his own labor as an artist, questioning while simultaneously profiting from the commercial marketplace and labor-saving mechanized processes of fossil modernity. I suppose that my own work follows the formula Aronoff gives; what is at stake in studying Melville studies is, for me as with the others, new ideas about the study of American literature as a discipline. But I want to stress the environmental dimension of these questions about US literature. Although the histories of literature and publishing are rarely examined as sites of environmental imagination, they do not fall outside the realm of environmental thought. Rockwell Kent and other artists and writers of the Melville Revival saw *Moby-Dick*, at least in part, as a text that helped them understand energy transition and fossil modernity.

ENERGY ARCHAEOLOGY IN THE MELVILLE REVIVAL

It was commonplace among the early critics of the Melville Revival to acknowledge the place of whale oil in the trajectory of energy history. Weaver and Mumford both repeat the old energy history chestnut that the discovery of oil at Drake's Well in Pennsylvania introduced a cheaper substitute for whale oil and dealt the mortal blow to the whale oil industry.[9] But a thread of energy critique runs even deeper throughout those early works of the Melville Revival. Calling on an older sense of the word "energy," Raymond Weaver in his biography in 1921 assessed Melville according to his high personal vitality and wrote that he "was blessed with a high degree of the resilience of youthful animal vigour" in his whaling days.[10] Weaver's praise for Melville's physical energy is of a piece with the way he and other conservative Melville critics held the author up as a specimen of idealized white masculinity, descended from the English, Dutch, and Scotch-Irish ancestors that Weaver called, in white supremacist language, "the best American stock."[11] In the rise of fossil modernity, as Daggett has argued, "energy" transformed from a word meaning personal vitality to one meaning the abstract units of work that fossil fuels made possible.[12] Approached through energy history, the approbation of Melville's energetic physicality sounds like a lament for the changing etymology of the word "energy" and the displacement of energy from a (white) man's body to the bodies of racially marked immigrants and industrial machines. But Weaver admits that even Melville's extraordinary vigor waned in the last part of his life, which Weaver calls the "long quietus." In Weaver's view, *Moby-Dick* received and expressed all Melville's youthful energy: "All [critics] wonderfully agree upon the elementary force of *Moby-Dick*, its vitality, its thrilling power."[13]

Lewis Mumford, in his biography in 1929, also wrote an energy-centered narrative of Melville's life, marked by a depletion of vital force after the completion of *Moby-Dick*. Mumford's energy language is even more technologically explicit than Weaver's when he compares Melville to a spent battery: "Melville was exhausted, exhausted and overwrought. In the prodigious orchestration of Moby-Dick, Melville had drained his energies. . . . Books like this are written out of health and energy, but they do not leave health and energy behind. . . . Moby-Dick had disintegrated him; by some interior electrolysis, its sanative salt was broken up into baneful chemical elements."[14] But Mumford's engagement of energy is even wider than this metaphor suggests.

Mumford's later work, *Technics and Civilization* (1934), offers a sweeping history of the world through technological and energy change, and the work has been called by some an originating work of the energy humanities.[15] Mumford evokes energy in the wide-ranging sense I use throughout *Rendered Obsolete*: energy in the old sense as vitality, but also as a word that denotes the technological and social changes wrought by expanding access to fossil fuels and a new political order around work. In *Technics and Civilization*, Mumford cites the development of the coal industry as the beginning a new era in human history, "Carboniferous Capitalism," anticipating later critiques of fossil capital and Anthropocene criticism.[16] Mumford describes the wide-ranging deleterious effects of fossil fuel extraction: "The animus of mining affected the entire economic and social organism. . . . The psychological results of carboniferous capitalism—the lowered morale, the expectation of getting something for nothing, the disregard for a balanced mode of production and consumption, the habituation to wreckage and debris as part of the normal human environment—all these results were plainly mischievous."[17]

Mumford had already begun developing his ideas about the relationship between energy and cultural production in the literary history *The Golden Day* (1926) and especially in his Melville biography. According to Mumford, Melville witnessed the world changed by the transformation Mumford calls "industrialism" in the Melville biography, and *Moby-Dick* was his testament of a world on the cusp of that change. Industrialism, according to Mumford in his Melville biography, was not only caused by new factory technology and industrial order but was incited by an acceleration of resource extraction and was itself the cause of widespread environmental destruction, as well as changes in life and work.

> Was this a triumph or a debacle, this coming of industrialism, this volcanic intrusion of new methods of living, new means of communication, new habits of work? When one thinks of the countrysides ran down, the forests that were wantonly depleted, the soils that were depleted, the towns that were jerry-built and burned and jerry-built again, the public lands that were thrown into the laps of speculators, the industrial population that was starved and depressed in dingy cities, one sees that there is no easy answer to this question; and certainly none of the economists has ever been impartial enough as accountant to tell whether the final result for civilization was a gain or a loss, and if so, how much was gained and how much lost, and where these things happened. But when Melville came back to America, industrialism was a value in itself: people encouraged it as the patrons of the Renaissance encouraged art, not doubting that activity was a great one, and made for a higher civilization.[18]

In the Melville biography, Mumford's account of the widespread economic and social changes wrought by industrialism already aligns with the idea of energy based in fossil fuels that he would develop in *Technics and Civilization*. Later in the biography, Mumford names "energy" and "work" even more explicitly as the sites of change in the mid-nineteenth century, and he frequently decries the way in which the new energy economy made work into a value in and of itself: "Where there is form and culture, there is true conservation of energy through the arts: where there is only energy, without end or form, the mechanism may be speeded up indefinitely without increasing anything except the waste and lost motion."[19] Mumford sees the changed world of America after the mid-nineteenth century as such a society "where there is only energy" and that suffers from ills he characterizes in the thermodynamic terms of "waste" and "lost motion." This deleterious new energy economy—"industrialism"—took hold in the United States in the 1840s, while Melville was away on his whaling voyage, so that he returned home to a changed world.

In Mumford's account, the sea serves as a byword for the world before industrialism, and it was Melville's experience at sea and his long association with it that gave him access to an older, better world: the "Golden Day."[20] Like Melville himself, Mumford assigned pride of place in the old seafaring culture to the commercial whaling industry that Melville chronicled in *Moby-Dick*. The whaling industry Melville chronicled seemed important to Mumford because it helped Melville cultivate the practical and intellectual parts of himself, but also because the whaling industry was at the cusp of its decline when Melville had joined it: "Melville had the singular fortune to pronounce a valedictory on many ways of life and scenes that were becoming extinct."[21] Ships and the sea also function metaphorically for Mumford and help him describe the enervating effect of industrialism on literature and art. Mumford describes Golden Day writers such as Melville who could not cope in the industrializing world as "battered wrecks" who had "lost the power to go to sea again; and instead of being beaten to pieces on the treacherous coast, merely rotted at the wharf."[22] Mumford had referents for old ships, even whaling ships, rotting at the wharf: he knew of the dilapidated ships in New Bedford harbor and of the *Charles W. Morgan* entombed in concrete.[23] Mumford figured the world that Melville and others were alienated from—the past world—as the sea. Many books and articles from the revival assess *Moby-Dick* great because it is the best in a subgenre of sea stories. This is how D. H. Lawrence, for example, classifies *Moby-Dick*: "It is an epic of the sea such as no man has equalled; and it is a book of exoteric symbolism of profound significance, and of considerable tiresomeness. But it is a great book, a very great book, the greatest book of the sea ever written."[24] I suspect that the sea is a byword or shortcut that stands for the bygone organic energy regime for many other critics in the Melville Revival, even if they did not share Mumford's explicit interest in technological change. As the Melville Revival wore on, fewer critics wrote about *Moby-Dick* as only a great sea novel, moving beyond both the subgenre of sea literature and a materialist interest in oceanic and energy history.

An important exception is Charles Olson, who, in the second wave of the Melville Revival, picked up the earlier critics' inquiries into *Moby-Dick* and energy.[25] In *Call Me Ishmael* (1947), Olson uses the word "machine" to describe the propulsive technology Americans have used to harness nature, which he reads as one of *Moby-Dick*'s major themes:

Americans still fancy themselves such democrats. But their triumphs are of the machine. It is the only master of space the average person ever knows, oxwheel to piston, muscle to jet. It gives trajectory.

To Melville it was not the will to be free but the will to overwhelm nature that lies at the bottom of us as individuals and a people. Ahab is no democrat. Moby-Dick, antagonist, is only king of natural force, resource.

I am interested in a Melville who decided sometime in 1850 to write a book about the whaling industry and what happened to a man in command of one of the most successful machines Americans had perfected up to that time—the whaleship.[26]

Olson casts whaling and *Moby-Dick* into a panoramic energy history, evoking the transition from organic to fossil fuel energy in the evolution of the "machine" from "oxwheel to piston, muscle to jet." Like Mumford, Olson casts a critical eye on extractivism in general and, in particular, on Ahab's extractivist drive: "the will to overwhelm nature" that he takes out on the whale Moby-Dick, "king of natural force, resource." But unlike Mumford, Olson finds more continuity than difference between the whaling world Melville chronicled and the world of pistons and jets. The whaling ship is just one more "machine" and, at that, the "most successful . . . Americans had perfected up to that time." For Mumford, *Moby-Dick* speaks from and about the Golden Day, whereas for Olson, *Moby-Dick* speaks to the fossil-modern future.

Rockwell Kent's *Moby-Dick* entered a field of criticism in the Melville Revival that was already much concerned with energy transition and the distance between the world Melville chronicled and the present day. Kent picked up many of the themes developed by Weaver and Mumford, and he anticipated the view of *Moby-Dick* that Charles Olson would put forward in the near future.

ROCKWELL KENT AND THE WOODEN WORLD: FORM

Kent's illustrated *Moby-Dick* was published in 1930 in two editions: an exclusive three-volume art book printed by Lakeside Press, which was an imprint of Chicago printer R. R. Donnelley, and a more accessible single-volume edition issued by Random House.[27] The three-volume Lakeside Press edition was limited to one thousand copies and was intended an advertising object through which Lakeside Press could flex its capacity as a fine book publisher. At Kent's urging, Lakeside contracted with Random House to print the Rockwell Kent illustrated edition in a smaller scale, a "pocket edition," that would be produced and sold en masse. The Random

House pocket edition became a best-seller in no small part on the basis of Rockwell Kent's reputation. In fact, it was Rockwell Kent whose name was on the front cover of the book—not Herman Melville's. Robert Frost wryly noted Kent's eclipse of Melville's fame:

> There is a story you may have forgotten
> About a whale....
> Oh, you mean Moby Dick
> By Rockwell Kent that everybody's reading.[28]

For the next few decades after 1930, *Moby-Dick* and even whaling history itself coursed through US culture in the black-and-white lines of Rockwell Kent's iconic illustrations.

Among the works of the early Melville Revival, Rockwell Kent's *Moby-Dick* offers the most complex inquiry into energy, work, and art. But these ideas must be teased out of the dark black lines of the Kent edition. Most critics approach Kent's illustrations as an interpretation of Melville's novel, searching for thematic continuities and discontinuities between the narrative and the images. Elizabeth Schultz writes that Kent's *Moby-Dick* illustrations "reveal and reinforce the novel's antitheses, paradoxes, and ambiguities [. . . and they] reminder the reader of the difficulties of discovering meaning in signs."[29] Art historian Angela Miller, for example, notes that Kent's black-and-white illustrations resonate with Ahab's worldview, which she characterizes as self-mythologizing and even, in light of Cold War anxieties that would characterize the novel's reception later in the twentieth century, totalitarian.[30] When I refer to Rockwell Kent's *Moby-Dick*, I refer to the form and content of those illustrations, to the integration of text and image in the final book product that Kent was so instrumental in designing, to the method by which Kent created the *Moby-Dick* illustrations, and to pieces in the larger body of Kent's artistic and literary work that provide necessary context for the *Moby-Dick* illustrations. It is through such a holistic approach to Kent's *Moby-Dick* that its energy critique becomes legible. Kent has been well known since his own day for his socialism, for his broad support of organized labor, and for his own part in representing laborers as heroic figures in his art and writing. In Kent's *Moby-Dick*, and in other associated works in which Kent represented the sea and maritime labor, the environmental dimension of Kent's labor politics begins to emerge.

Rockwell Kent's famous illustrations for *Moby-Dick* cast Herman Melville's novel into a world made out of wood. Consider Kent's full-page portrait of Captain Ahab, who enlists the crew of his whaling ship to pursue vengeance on the eponymous white sperm whale who took off his leg. In Kent's portrait, Ahab stands on the wooden deck of his whaling ship and gazes out to sea. Everything in the ship is made of long, almost parallel lines like the grain in wood. The illustration delineates individual wooden planks on the ship's deck; on the side railing, two small curls suggest the circular head of a bolt or nail and evoke the carpentry that made the ship. The wood aesthetic attaches to everything in the image. Ahab's watch coat is composed of heavy horizontal lines that, in large part, run parallel to the lines of the wooden ship he stands on. The coat is fastened like the wooden railing behind him, drawn together by four circular buttons that echo the bolt heads. The sea horizon is constituted by lines stacked so tightly that the ocean appears almost black. Not even the clouds in Kent's linear illustration are permitted to be soft or cottony; they are given dimension through rigid dots and dashes. Ahab's face is striated, too: shadows falling across the deeply lined face are marked with short, black lines. Ahab's stern face appears carved in wood, the angry and sorrowful lines in his face engraved as deeply as his hollowed-out eyes and down-turned mouth. Ahab's ivory prosthesis is edged in fine-toothed shadows, the only unlined object in the scene.

Ahab's portrait, like each of Kent's illustrations for *Moby-Dick*, appears to be a wood engraving, and the match between the subject and the medium could hardly be better suited. The lines, textures, palette, and formal echoes in Kent's portrait of Captain Ahab all point to the wooden block upon which an engraver would have cut the stylized image in reverse. One of the most consistent aspects of Kent's illustrations for *Moby-Dick*—an aspect of the work fully evident in the portrait of Ahab—is that they all appear to be woodblock prints, a practice of relief printing that print historian Richard Benson defines as "printing from the high parts of some surface.... Parts of the block's surface have been cut away, and leave no mark, appearing in the image as white. The high parts of the block have not been cut away, and they print black." Another important characteristic of this medium is that it is "fundamentally linear" and that relief printing "has no mechanism for making black ink print gray."[31] By Benson's definition, Kent's portrait of Ahab appears to be a woodcut or wood engraving. The illustration does not exhibit the fine cross-hatching more common in intaglio printing, as in copperplate engraving or etching. Kent's illustrations are rendered in thick lines, parallel and horizontal like woodgrain itself. Ahab even appears to be an engraver's avatar, the sharp point of his ivory leg standing on the wooden deck so like the engraver's burin on a wooden block. Wood, strong and rigid, is symbolic of Ahab's obsession with the white whale.

Rockwell Kent cast Ahab into a world made of wood: the ship, the man, and even the sea and clouds are rendered in rigid lines that evoke a woodblock print. Rockwell Kent, illustration for *Moby-Dick*, chapter 46. Original medium: ink drawing on paper. Rights courtesy of Plattsburgh State Art Museum, State University of New York, Rockwell Kent Collection, Bequest of Sally Kent Gordon. All rights reserved.

Kent's emphasis on the wooden world of *Moby-Dick* anticipates the argument that Lewis Mumford would make in *Technics and Civilization*. There, Mumford divided the history of the world into technological regimes that somewhat overlap each other: the "eotechnic," running through the mid-eighteenth century, the "paleotechnic," running from the mid-eighteenth century through the present, and the "neotechnic" in the twentieth century beyond. "Wood," wrote Mumford, "was the universal material of the eotechnic economy" (whereas "paleotechnic industry rested on the mine").[32] Mumford's eras, each defined by a different technological energy regime, resemble the efforts of later energy humanists to periodize literature and culture by energy source, such as Patricia Yaeger, who proposed to "sort texts according to the energy sources that made them possible."[33] With those wooden illustrations, Rockwell Kent's *Moby-Dick* presents the whaling world of *Moby-Dick* as a world made from wood, an exemplar of the eotechnic. And indeed wood made up the infrastructure of the nineteenth-century whaling industry, from the wooden ships to the wooden harpoons, from the wood that fed the blubber-rendering fires to the wooden casks filled with whale oil. Mumford made an argument about the bygone world through words, Rockwell Kent through the form of the woodblock print.

An accomplished printmaker, Kent well understood the practices of woodblock relief carving and the formal characteristics of finished prints. After beginning his career as a painter, Kent made his first major wood engraving in 1919 at the behest of friend Carl Zigrosser, a printer and gallerist, who gave him a hand press and taught him how to use it.[34] Kent created woodcuts and wood engravings throughout the rest of his career. He worked self-consciously within the tradition of print history, even adopting the pseudonym "Hogarth, Jr." for the satirical illustrations he made at *Vanity Fair*.[35] Advertising art would be one of Kent's main sources of financial support, and his woodblock-style illustrations were used to sell everything from fine jewelry to yachts. While Kent created the visual language in which so many companies spoke to their customers, he simultaneously branded himself with the woodblock style. Reviewing nearly two decades of wood engraving in the United States, in 1936 critic and fellow wood engraver Clare Leighton credited Kent with popularizing the medium: "The best known of American graphic artists is Rockwell Kent. ... It is probably he who has made the general American public aware of the wood block."[36]

Kent's woodblock style became associated with the wider field of book publishing, partly on account of those popular illustrations and partly because he designed enduring colophons for several publishing houses,

Rockwell Kent deployed the woodblock style in colophon designs for many publishing houses, including Random House and Viking Press (*pictured here*). Kent's signature designs cemented the relationship between classic US literature, the contemporary publishing industry, and the woodblock style. From Rockwell Kent, *The Bookplates and Marks of Rockwell Kent* (Pynson Printers for Random House, 1929). Rights courtesy of Plattsburgh State Art Museum, State University of New York, Rockwell Kent Collection, Bequest of Sally Kent Gordon. All rights reserved. Image courtesy Rare Book & Manuscript Library, University of Illinois at Urbana–Champaign.

including Viking Press and Random House. Perhaps seeking visual coherence with the texts they decorate, the sparse black shapes and attenuated bodies of Kent's woodblock colophons resemble the rounded forms and thin serif feet of the Roman typefaces favored by contemporary publishers. To go a step further, the smooth horizontal lines of the woodblock style rhyme with the horizontal lines of text that surround them on the page. Schultz notes that the horizontal lines of Kent's illustrations "move the reader's eye back to Melville's text." I argue that it is not just Melville's text that is evoked through the horizonal lines of Kent's woodblock style but all printed literature.[37] The formal harmonies between woodblock image and printed text underpin Kent's woodblock-style colophons for the publishing houses of New York and answer a need born specifically of the early twentieth-century literary scene to define US literature. As critics openly sparred over which texts constituted the American literary canon, publishers responded by rapidly printing new editions of freshly anointed classics. Kent's wood-engraved colophons offered American publishing a distinctive visual vernacular for that new type of book, the "American classic." Kent's woodblock style was a formal mechanism by which trade publishers claimed authority on classic literature.

Wood engraving invites viewers to reflect on labor, on the history of sequential tasks required to produce a printed image: drawing or sketching, engraving, printing on a press. As Kristin Bluemel remarks, "Wood engraved illustrated texts, with their dependence on contrasts of depth on the wood block to achieve contrasts of meaning on the flat surface of the page . . . invite not only formal or 'surface' analysis of image and word, but also 'depth' analysis of historical and social relations."[38] It is easy to imagine how the act of engraving on wood would appeal to Kent's leftist political views; the American left often expressed its political aspirations for unalienated work in images that glorified manual labor.

And, indeed, in 1937 Kent created an iconic wood engraving, *Workers of the World, Unite!*, that was used as a cover illustration for the Marxist magazine *New Masses*.[39] The wood medium resonates with the subject of the engraving—a colossal laborer raising a spade against foes just outside of the image. The worker reverses the biblical injunction, making his plowshare into a sword. "Workers of the World, Unite!" links the worker's labor and his heroic revolt with the medium of wood engraving. Drawing a burin precisely and at length through a rigid block of wood is a taxing physical process, a practice of penetration and exertion, like the work of the heroic laborer.

In aligning the practice of wood engraving with manual labor—and specifically agricultural labor—Kent's woodblock work fulfills the theory of engraving that the Victorian polymath John Ruskin offered a half-century earlier. Ruskin defined engraving by the labor that carves the block rather than by some formal characteristic of the finished print: "To engrave is, in final strictness, 'to decorate a surface with furrows.' . . . A ploughed field is the purest type of such art; and is, on hilly land, an exquisite piece of decoration. Therefore it will follow that engraving distinguishes itself from ordinary drawing by greater need of muscular effort." Ruskin actually called engraving tools by the name "ploughshare" and compared the act of cutting wood or metal to plowing a field, evoking, Mumford-like, the act of wood engraving with the organic energy regimes of human and animal muscle on a farm.[40] Likening wood and metal engraving to the elemental labor of plowing a field, Ruskin established a metaphor relevant to the idealization of manual labor embedded in Kent's nostalgic illustrations for *Moby-Dick*.

ROCKWELL KENT AND THE WOODEN WORLD: METHOD

It is necessary to subdivide my discussion of the *form* of Kent's *Moby-Dick* illustrations from the *method* by which he produced them, because Kent's illustrations for *Moby-Dick* are *not* wood engravings or woodcuts.[41] They are ink drawings that Kent created using a brush and ink and sent off to the printer in Chicago, where they were photographed and—depending on the demands of each illustration—hand-engraved, transferred by electrotyping onto a metal plate, or both, by a team of engravers at R. R. Donnelley. Kent did indeed produce a large body of wood engravings and woodcuts, but most of the illustrations he created for *Moby-Dick* and other books were executed in pen and ink and surrendered to the publisher for a thoroughly modern form of translation into print. Kent never intended to pass any of his drawings off as something they were not, but some critics and collectors have since misidentified Kent's ink drawings as woodcuts or engravings, sometimes obfuscating the medium of these works with the catch-all term "print."[42] These errors and omissions have persisted through catalogs and scholarship, but the confusion about the medium of Kent's work is important for other reasons: that confusion signals anxiety about how to understand a work made in one medium (ink drawing) that looks as if it has been produced in another medium (wood engraving or woodcut).

Why did Kent mimic the look of a woodblock print in another medium? I read his woodblock print simulations as "skeuomorphs," artifacts with design features that resemble the essential characteristics of another medium and an earlier technological process. Skeuomorphs perform the furtive cultural work of nostalgia, and Kent's illustrations for the 1930 reprint of *Moby-Dick* occupied a nostalgic nexus that is most obvious in its contribution to the American print revival and its service to literary canon formation. Kent's *Moby-Dick* mediates energy transition by calling on all of the political and technological associations of wood, but his skeuomorphic illustrations explode the very idea of "medium" by requiring a bifurcated approach to form and method.[43]

Kent's woodblock prints and the skeuomorphic drawings that replicated their "woodblock style" reveal his nostalgia for unalienated, pre-fossil-modern labor. Critic Svetlana Boym defined nostalgia as "a longing for a home that no longer exists or has never existed. Nostalgia is a sentiment of loss and displacement, but it is also a romance with one's own fantasy."[44] Kent's woodblock-style skeuomorphs produce the effect of print nostalgia, in the sense of Boym's understanding of nostalgia as a fantasy or a utopic vision rather than as a faithful representation of history. It is impossible for Kent's skeuomorphs, or for any act of history, to offer unmediated access to the past; instead, the skeuomorphs record Kent's imagination of the pre-fossil-modern past and his longing to access that vision.

In Kent's time, the woodblock style represented print's past. New technologies for reproducing images were introduced around the turn of the twentieth century. Through the late nineteenth century, book and periodical publishers widely used wood engravings to reproduce images. But in the late nineteenth century, developments in photochemical processes and half-tone printing brought wood engraving nearly to an end. When wood and other forms of engraving ceased to be necessary to reproduce images, the practice did not disappear but migrated to the sphere of artistic expression. Leighton locates the wood engraving revival—she calls it the "rebirth of the woodcut" and the "renaissance"—to the early 1920s, when artists who took up the technique exploited the medium's woody reflexivity. By 1936, with the publication of her review *Wood Engraving of the 1930's*, she writes that "wood engraving has now emerged from all this adolescent behaviour to a truly mature existence."⁴⁵ In spite of these claims of woodcutting's "revival" and "rebirth" as an artistic practice in the early twentieth century, it is fair to say that wood engraving never fully died in the short interim—rather, that artists took up the medium by choice rather than from technological necessity.⁴⁶

The woodblock style made *Moby-Dick* look like the classic it had become: a valuable work that spoke from the past. It was not only in Kent's woodblock-style illustrations that he expressed an ambivalent affinity for the old-fashioned. He writes in his memoir *This Is My Own* (1940) that he was trained to revere old forms, and when he set about to build his house in Upstate New York, he turned to the education that had taught him to consult the past:

> Wasn't I a product of the Columbia University School of Architecture, 1904, and conditioned by its teaching to believe that railroad stations should be like Caracalla's baths, banks like Roman temples, churches like Chartres or St. Paul's, city houses after Bramante and Sansovino, country houses like (here we were liberals!) French châteaux, or Tudor or Georgian mansions, or—well, like anything quite old and quite unlike ourselves? And wasn't I still young enough, at the subsequent awakening of American taste to the glory that had been America two centuries ago, to share in that awakening? And almost young enough a few years later to half accept the thought that Victorian godawfulness was beautiful? I—we, in fact, together with most of our generation—were quite old enough twelve years ago to have accepted as basic to good taste the principle that as new things made to look old were good, old things were best.⁴⁷

Kent lets slip a kernel of embarrassment in parentheses—"(here we were liberals!)"—when he admits that he looked to the aristocratic past, rather than to the past of ordinary working people, in making designs for modern buildings. The training and principles that guided Kent in designing his house speak to the principles that guided his hand while he drew illustrations for *Moby-Dick*. The dictum of taste that Kent admits here—that "new things made to look old were good"—offers a concise definition of skeuomorphism without naming it.

Nostalgia for unmechanized, physical labor was one of the most durable fantasies evoked by Kent's skeuomorphs. For example, he created a series of memorable drawings for *Wilderness* (1920), the account of his long sojourn in Alaska. Many of these drawings are playfully self-referential; they look like wood engravings about wood. Though looser than the strict, stylized

Kent made this illustration about woodcutting, *Fire Wood*, in the woodblock style, but he executed the illustration in pen and ink. As a portrayal of labor that did not happen, the illustration is an apt metaphor for skeuomorphism. Illustration for Rockwell Kent, *Wilderness: A Journal of Quiet Adventure in Alaska*, 1920 (Middletown, CT: Wesleyan University Press, 1996), 13. Original medium: ink drawing on paper. Rights courtesy of Plattsburgh State Art Museum, State University of New York, Rockwell Kent Collection, Bequest of Sally Kent Gordon. All rights reserved.

Moby-Dick illustrations, his drawings for *Wilderness* often depict figures engaged in wood-cutting or whittling; in these images Kent gestures even more blatantly at the wood-cutting labor he did not perform in making the images. A drawing called *Fire Wood* shows a man and a young boy at either end of a two-handled saw, the log they are cutting at the center of the image. Above the log, the sun sets or rises at the mountainous horizon, and black lines radiate outward from the center all the way to the edges of the image. The whole scene is a circle, emanating outward from the cut log at the center. But, as no wood was cut in the making of this image, *Fire Wood* offers an apt metaphor for skeuomorphism: it portrays labor that did not happen. By using the look—but not the fact—of the woodblock print to signal old forms of labor, Kent's nostalgic skeuomorphs commemorate a bygone ideal of labor rather than its actual execution. Kent's prints and illustrations transmit his discomfort with the economic and technological world he inhabits and, at the same time, his complicity with it.

SKEUOMORPHISM

The word "skeuomorph" derives from the Greek words meaning "vessel" or "implement" and "form," and according to the *Oxford English Dictionary*, a skeuomorph is "an object or feature copying the design of a similar artefact in another material."[48] A skeuomorph is a nostalgic type of remediation, a simulacrum that gestures specifically toward the past. Typically, the medium to which a skeuomorph refers is an obsolete or old medium, a decorative vestige that cites past techniques and technologies. Or, as media theorist Dan O'Hara has written, the general principle of skeuomorphism is that "where a function is rendered obsolete, its residual traces become ornament."[49] Wood, like whale oil, has by the time Kent created his illustrations for *Moby-Dick* been rendered obsolete.

The nostalgic dimension of skeuomorphism is so familiar to us today that it seems almost natural. The digital world appears in many places as a loving monument to the analog: digital photo filters confer Kodachrome warmth or sepia tones to cold digital images, and films and television shows shot on digital cameras are edited in postproduction so that images appear scratched, grainy, and otherwise shot through with the imperfections of analog film.[50] Media studies scholar Dominik Schrey calls the trend "analog nostalgia," connecting the fragmentary and damaged aesthetic of skeuomorphic image-making to the fashion for constructing follies—artificial ruins—in seventeenth- and eighteenth-century Europe.[51]

From a disinterested scholarly point of view, skeuomorphism appears as a wordless form of media archaeology, a media studies approach to artifacts and practices that did not circulate under the name "media" in their time but that operated in their own time in a way that resembles "media" in ours.[52] Media archaeology offers an approach to understanding what Erkki Huhtamo has called the "neglected, misrepresented, and/or suppressed aspects of both media's past(s) and their present."[53] Kent's skeuomorphic ink drawings reveal the labor practices and ideologies behind artisanal woodblock printing and alienated, multistep drawing and electrotyping. And Kent's skeuomorphic drawings forge, too, a link between media archaeology and energy archaeology; they demonstrate in content and creation the link between old and new energy regimes.

From the point of view of product designers and marketers, skeuomorphism serves a useful function by easing consumers into new products with the familiar forms of the past. Here in the twenty-first century, the computer or smartphone offers a rich archive of skeuomorphism. Steve Jobs and Apple designers deliberately deployed skeuomorphism as a strategy to bring consumers into a closer relationship with Apple products.

Digital "files" are stored in digital "folders" and disposed of in "recycle bins," all represented in digital space by their counterparts in the analog, material world of the typical office—which, not incidentally, Apple hopes that its products might inalterably change and even obviate.[54] Apple began moving away from skeuomorphism around 2013 under the direction of Apple Chief Design Officer Jonathan Ive, who has tried to design skeuomorphism out of recent editions of Apple's operating system in favor of what he calls "flat" design based on the principles of modernism. But Apple has returned to skeuomorphic design with Apple watches, which display the image of an analog watch face.[55] It is poignant that Apple has returned to skeuomorphism with watches, because skeuomorphic nostalgia is inherently about the loss of a worker's authority over their time and tools. In Rockwell Kent's day and in our own, skeuomorphs register our anxiety about the alienation of labor, as well as the way that work can engulf other parts of our life—anxiety, in short, about how we spend our time.

THE POLITICS OF THE WOODBLOCK

It is not necessary to rely solely on Kent's engravings to understand his views on labor. Kent wrote volubly—if often paradoxically—about labor and specifically about the political dimensions of work. Kent's political opinions ranged across a wide spectrum of leftist ideology: he was an outspoken and lifelong socialist and a member and one-time president of the International Workers Order, a communist organization. He was hauled before Senator Joseph McCarthy's Permanent Subcommittee on Investigations in 1953 to discuss his affiliation with communist organizations. Even after most American leftists lost faith in the Soviet experiment in communism, Kent remained avidly interested in the USSR. In 1960, he donated a huge body of his work to the Soviet Union, and in 1967, he won the Lenin Peace Prize, the Soviet answer to the Nobel.[56] Part of the leftist critique of capitalism one might expect to find does, in fact, persist in Kent's art. His nostalgia is a reaction to modern capitalism and, specifically, to how it estranges the laborer from the materials, product, and meaning of his or her labor. Kent's fantasy of the past resembles, at least in part, the old Marxian dream of unalienated labor.

But Kent's leftist values do not always consistently apply to his own work. His theory of printmaking was financial, fully complicit with the logic of the capitalist marketplace: wood engravings were good because customers could buy original art at low prices, and artists could earn a living by producing prints. In 1934, Kent published a small pamphlet called "How I Make a Wood Cut" for a series of museum guides called *Enjoy Your*

Museum. In that pamphlet Kent claims that the power of printing was in what he called its "democratic" power to make original art accessible to a wide audience:

> [Printing] is a form of picture making in which the final printed product is in every sense the true "original" of the design; an "original" toward the making of which the sketch, the plate or block or stone, are only steps of a process directed toward one end: the perfect printed thing. And since by such a means of picture making there may be produced at last a number of identical "originals," print making may be truly termed a democratic art.[57]

Kent's discussion, here, of the democratic art obscures the steps of the process by which those multiple originals are produced (sketch, block, and so on) and focuses on the "perfect printed thing," which, in fact, becomes several identical things: multiple works of art liberated from small, elite enclaves and made available to a wide group. The politics of "multiple originals" are appealing to a leftist such as Kent, who believed that art belonged to everyone.[58]

Elsewhere, Kent deploys the same term—democratic—to advocate for artists to make prints; as well as making prints more accessible to consumers, a press allows artists to produce art—and make money—on a larger scale: "One may say that, as far as art is concerned, all processes are equally difficult or simple. Each yields its proper and different result. And that the artist chooses at all to produce prints rather than drawings is in consideration both of the nature of the result and of the economic necessity of making originals from one design. Printmaking is a democratic art."[59] In evoking the "economic necessity" of making prints, Kent acknowledges the benefits for artists of an economy of scale: printmaking brings in more money for less labor than other forms of art. Kent writes in a straightforward way about "economic necessity," but in his milieu, the question of whether and how leftist artists should produce commercial art was fraught with political implications. John Sloan, like Kent, was a prolific commercial illustrator and a close friend of Kent's; in November 1910, Kent and his wife, Kathleen, rented an apartment in the same building as Sloan. In his diary entry for June 6, 1910 (just a few months before Kent moved in downstairs), Sloan recounts an argument with the art editor Edward Wyatt Davis, who had taken on extra work producing advertisement copy for Borden's Milk Company. "[Davis] is prospering. Says that he believes in all the ideas of Socialism but has made up his mind to get money if he can. Deliberately shutting out what he really knows is true. In other words, as the 'system' still stands to get what he can of the spoils. This does not seem wrong to me, it merely is the position of one who decides not to take up the cause and fight against the present."[60] Kent's attitude toward commercial art is less resigned than Sloan's depiction of Wyatt; he seems to have believed that he could remain an ideologically pure anticapitalist even while prospering in commercial art.

Kent's perspectives on art and commercialism are paradoxical: art can and should resist capitalism oppression and alienation, and it should make as much money as possible. Kent's method for sustaining this paradox was to compare commercial art to extraction in an extended metaphor. In his memoir *This Is My Own*, Kent calls New York the "gold camp": "It was a miner's camp, New York, all through the twenties, a place to work in, save—if one was wise; get rich enough, pack up, clear out."[61] Through metaphor, commercial art became strenuous labor. Kent lived in New York on and off throughout the first two decades of the twentieth century. For many of those years, Kent worked on and off as a draftsman, a "renderer" in the firm of Ewing and Chappell: a job he likened in his memoir to working on a "chain gang," another metaphor of grinding physical labor for the quiet yet physically enervating work that the artist performed at a desk. Kent produced commercial art on commission on the side until he felt secure enough in his commissions to resign in 1912 and freelance full-time.[62] It is telling that Kent resorted to metaphors of extractive labor—gold mining and prison labor—to explain how he made money. While Kent critiqued the aesthetics of fossil-modern labor, he instrumentalized and ultimately profited from its violence without ever confronting it.

ROCKWELL KENT, INC.

The paradox of Kent's views on commercial art practice in light of his leftist politics has made him vulnerable to enduring charges of hypocrisy. Corey Ford, a contemporary satirist, pilloried Kent as a greedy panderer. Ford was a prodigious satirist in the style of the Algonquin Round Table; he published columns in *Vanity Fair* and the *New Yorker*, in addition to book-length parodies and collections. Ford wrote satire-as-literary criticism and took special aim at writers like Kent who flaunted their quest for authentic experience.[63] Writing under a frequently used pseudonym, John Riddell, Ford published a parody of *N by E*, Kent's narrative of his sailing voyage to Greenland. *N by E* was heavily illustrated with Kent's wood engravings and woodblock-style drawings. Ford parodied the book in a short essay called "N by G," published in the collection *In the Worst Possible Taste* (1932). Ford aped Kent's exclamatory prose style and his wood engravings by offering the fictional autobiographical voyage narrative of one "Rockwell Riddell."[64]

In a prefatory note, Ford (in Riddell's voice) names Kent, explaining that Rockwell Riddell's voyage was inspired by "a commercial illustrator named ROCKWELL KENT, INC."[65] The snarky "Inc." behind Kent's name serves as a concise critique of the way Kent embraced commercial art, but it also references a real event in Kent's life. Fully embracing the commercialization of art practice, Kent incorporated himself in 1919 and sold shares. Kent even created a shareholder's certificate—in the hallmark woodblock style, naturally.[66] From an avowed socialist like Kent, the incorporation might have been a political stunt that illustrated the complicity of artists with market ideology—but Kent needed the money. Kent's literal self-incorporation was irresistible to Ford, and throughout "N by G," he tacks on the "Inc." suffix to words throughout the story: "Art, Inc." and "Adventure, Inc." as well as "Rockwell Riddell, Inc." Ford's narrative ends nihilistically and solipsistically, with Rockwell Riddell, naked, running north into the icebound Greenland wilderness: "Everywhere there is nothing but Art; and me, in the middle. I have found the infiniteness of Infinity. Now I belong to Oblivion. Now I am one with Nothing. I am Nothing, Inc."[67]

Ford located Kent's opportunism in the black-and-white woodblock style itself, which was broadly imitated in an illustration accompanying Ford's essay. The illustration by Miguel Covarrubias is instantly recognizable as a Kent parody, featuring the same horizontal lines that scaffold his woodblock-style drawings. Ford writes in the caption: "Surrounded by one of his familiar black-and-white landscapes—complete with geometric icebergs, angular nudes and a slide-rule sunset, its rays numbered carefully from 10 to $10,000.00—Rockwell stands created at last in his own image."[68] Ford draws on the diction of rationality ("geometric," "angular," and "slide-rule") to describe Kent's illustrations. In this light, the linearity and the solidity of Kent's drawings—the same qualities that make them resemble woodcuts—bespeak their conformity to the rationalizations of the capitalist marketplace. A compass rose in the upper right-hand corner of Covarrubias's illustration mislabels the cardinal directions, with "S" at the eastern point and "E" (printed in reverse) at the western. The stacked horizontal lines that ladder the page are indeed numbered, as if to indicate their unit price. The numbers start with "10" at the top and continue in increments of 10 until they jump, greedily, to increments of 50, and then to 100, and then to 1,000 until they arrive at 10,000, changed to $10,000 in Ford's caption. The figure at the center of the illustration stands with legs spread wide and arms raised in distinctly Kentian exultation.

Time and again, Kent returned to the "gold camp" and got rich enough—although likely never so rich as critics like Ford might have imagined. In

Miguel Covarrubias's illustration *Rockwell Kent* is both a portrait of Rockwell Kent and an instantly recognizable parody of Kent's woodblock style. From John Riddell, *In the Worst Possible Taste* (C. Scribner's Sons, 1932). Printed book illustration © María Elena Rico Covarrubias. Courtesy Rare Book & Manuscript Library, University of Illinois at Urbana–Champaign.

fact, Kent's success as a commercial illustrator might have been one of the reasons that he stopped producing wood engravings and turned, instead, to pen-and-ink drawings in the style of wood engravings. In an account published in 1936, Clare Leighton reflects on Kent's possible motives for producing drawings rather than engravings: "In America some years back, Rockwell Kent was so overwhelmingly the fashion that firms were tumbling over each other to procure him. This slackened off during the slump, but not before Rockwell Kent had realised that it was his style that was wanted more than the fact of his engraving on wood; and wisely he decided to work in pen and ink."[69] Leighton hypothesizes that market opportunity drove Kent to work in pen and ink rather than on a wood block, so that he might diminish his labor and earn as much profit as possible while he had the chance, but she, like so many others, elides the difference between the woodblock and the woodblock style.

Woodblock or woodblock style—it apparently did not matter to Kent, for whom commercial art was just a way of making money, or even to viewers who enjoyed woodblock style irrespective of the practice that produced it. And the difference between method and form did not seem to matter to critics like Ford. Like Kent, they were more interested in what the woodblock style symbolized, rather than in the real thing itself. As fellow artists and workers in a fossil-modern world that distanced them from organic energy sources, they could afford not to care.

ROCKWELL KENT AND THE SEA: CONTENT

Rockwell Kent is perhaps the signal example of the Melville Revival figure I described above: the interpreter of *Moby-Dick* for whom the novel is great because it is the best sea story, and the thinker for whom the sea serves as shorthand for a bygone energy regime and the forms of work and life that it made possible. It is significant that Kent's work so often points to wooden infrastructures, tools, and styles, but it is just as significant that the subject of his work is so often the maritime world. The skill and exertion of sailing a wooden boat is the subject of the woodblock-style skeuomorphic drawings that Kent made to illustrate his sea stories, including *Moby-Dick*, and the narratives he wrote about his own sailing adventures: *Wilderness: A Journal of Quiet Adventure in Alaska* (1920), *Voyaging Southward from the Strait of Magellan* (1924), and *N by E* (1930). The illustrations for these works are always drawn to resemble woodcuts or wood engravings. Kent's enormous corpus of paintings, engravings, drawings, and writings all reveal that his fantasy of authenticity and honorable work is deeply rooted in place, or in a type of place. Kent yearned for, traveled to, and made images of the world's cold, forbidding, and remote islands and coastlines: Maine, Newfoundland, Greenland, Alaska, Tierra del Fuego.

Kent was part of a wide cohort in his idealization of Maine and the maritime world as sites where the past lay closer to the surface of everyday life. Among New York realists and modernists from the Ashcan School and Alfred Stieglitz circle, Maine became shorthand for labor and authentic experience, drawing artists including Robert Henri, George Bellows, and especially Marsden Hartley, a Maine native who self-consciously aligned himself with the state. Henri, Kent's teacher, introduced Kent to Monhegan Island. As Donna Cassidy has argued, the idea of Maine among tourists and modernist artists alike lay at the intersection of two fantasies: the fantasy of New England, an authentically American region with a strong connection to the past, and the fantasy of Maine residents as "North Atlantic folk," who were "antimodern, hardworking, rugged individuals."[70] The place of Maine in the twentieth-century artistic/touristic imaginary is not unlike that of Nantucket in the late nineteenth century, as discussed in chapter 2. Twentieth-century Maine, like nineteenth-century Nantucket, was revered for its quaintness: for the uncanny persistence—even if only in ruins—of the wooden sailing world and other forms of preindustrial labor. Nantucket was made accessible to quaintness hunters in the nineteenth century by the expansion of fossil-modern transportation networks such as rail and steamship lines. Maine, more remote from East Coast metropolitan centers, became newly accessible to tourists in the early twentieth century owing to the proliferation of cars, highways, and automobile tourism: fossil modernity producing new forms and aesthetics of quaintness through tourism.

Kent's Maine was Monhegan Island, a tiny island located twelve miles off the coast. The artist wrote that he discovered "real" work for the first time on Monhegan, where he lived on and off from around 1905 to 1910. The island was then (as it is now) home to a small year-round community of fishermen and an enclave of artists who came every summer to paint the island's lighthouse, sea views, and rocky headlands. On Monhegan, Kent consolidated the vibrant, realistic style that would define his work across many media and his preoccupation with maritime labor. Monhegan was also the place, according to Kent's memoir *It's Me O Lord* (1955), where encounters with fishermen and other laborers first awakened his class consciousness. Kent described watching the Monhegan fisherman from a cliff and experiencing a wave of shame at his own literally and figuratively elevated position of observation:

God, how I envied them their power to row! To pull their heavy traps! I'd seen my own thin wrists, my artist's hands. As though for the first time I saw my work in true perspective and felt its triviality. It is easier to do a job of work, said Oscar Wilde (I think he said, to dig a ditch), than to write about it. What rot! Why, Oscar never worked in all his life.

With these thoughts, this envy, agitating me, those social and political convictions which had hitherto existed as figments of my mind and heart began to acquire substance. Not just from Whitehead, but from empyrian heights had I looked down on life, down at the masses, and not with them, up.... What would I do, I thought, should working men not work for me? The answer shamed me.[71]

Kent tried to make himself into one of those tough Monhegan laborers, working for five years on the island as a well driller, lobsterman, longshoreman, and house builder. The analogue in paint for Kent's self-described political awakening is his painting *Toilers of the Sea*, the title taken from the Victor Hugo novel *Les travailleurs de la mer* (*Toilers of the Sea*, 1866). It shows fishermen hauling lobster traps in two small dories pitched precariously on rough seas near the spot described above. Kent admired Monhegan fishermen and the world they inhabited, too, because they appeared old-fashioned: "The fish-houses, evoking in their dilapidation those sad thoughts on the passage of time and the transitoriness of all things human so dear to the artistic soul; and the *people*, those hardy fisherfolk, those men garbed in their sea boots and their black or yellow oil skins, those horny-handed sons of toil."[72] The fishermen and fishing infrastructure at Monhegan were "dear to [Kent's] artistic soul" because they were dilapidated, because—like ruins in the Romantic imagination—they incited a pleasant melancholy. The fishermen represented not just work but old-fashioned work: work in which the laborer produced energy through his own muscle power to extract fish from the sea.

Yet Kent sustained himself by selling the image of those "sons of toil" to people who did not and could not perform this kind of work. Kent repressed fossil modernity with astonishing flamboyance in a series of advertising engravings he made in 1930 and 1931 for the American Car and Foundry Company, a railroad car and automobile parts manufacturer. Kent's work advertised the company's new side venture in luxury motor yachts.[73] The wood engravings feature stylized moonlit nude figures reposing on sailboats, an occasional mast fragment or rope ladder standing in for the figure of the whole boat. This time, Kent evokes the oceanic trinity

Kent's painting *Toilers of the Sea* memorializes his humbling encounter with fishermen on Monhegan Island, Maine. Courtesy of the New Britain Museum of American Art, Charles F. Smith Fund, 1944.01.

that characterizes his *Moby-Dick* illustrations—man, boat, sea—to find the sublime dimension of sport leisure at sea. Critic Jonathan Weinberg has explored the radical dimension of the engravings, noting the homoeroticism of the American Car and Foundry images and their resonances with the work of William Blake and Walt Whitman: "Nudity suggests freedom from bourgeois constraints and the possibility of a reconnection within the world of nature."[74]

But the American Car and Foundry yachts advertised in these images were the very definition of bourgeois consumerism. Moreover, they looked and operated absolutely nothing like the boats in Kent's engravings. The boats in Kent's images never appear in full view, but his nudes pose against

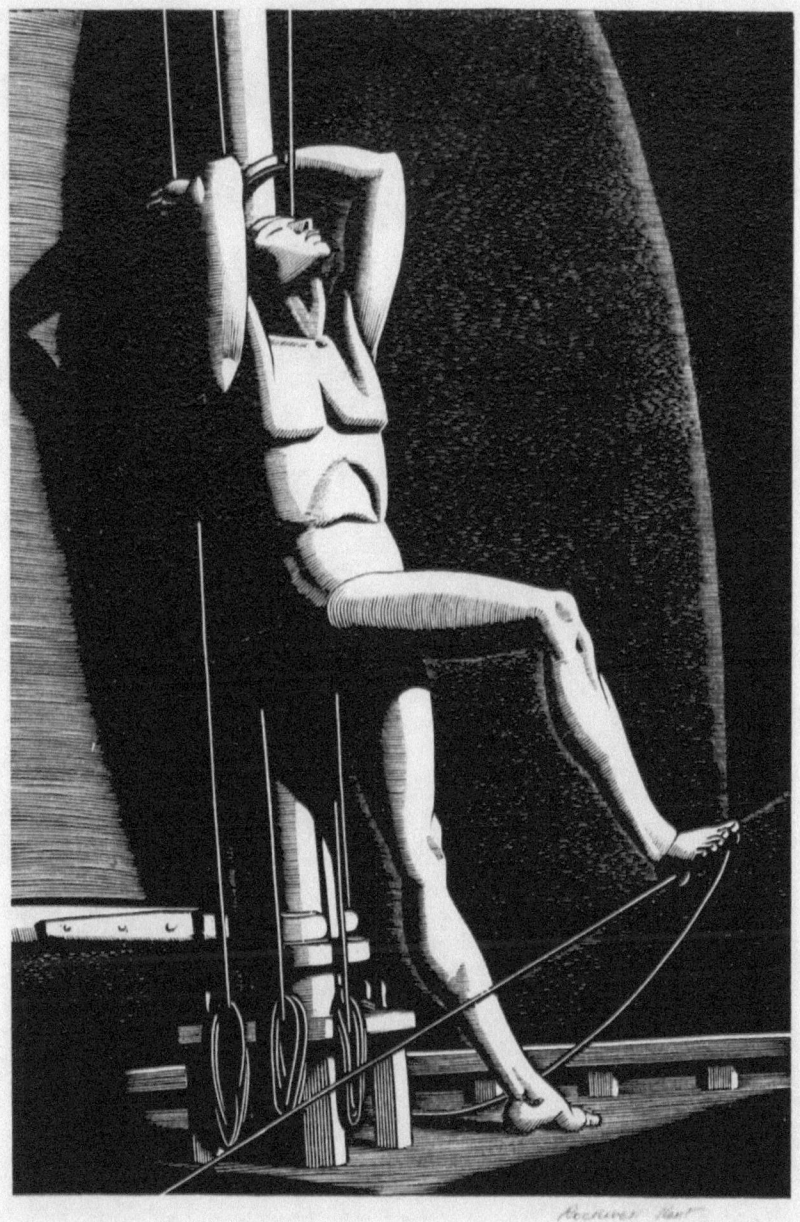

Kent's nostalgic use of wooden boats and the woodblock style persisted in a series of advertisements he made for the American Car and Foundry Company. These prints actually were wood engravings made by Kent, but the boats they advertised were for the American Car and Foundry Company's new line of motor yachts. Rockwell Kent, *Night Watch*, 1929, wood engraving, platemark: 20 × 13.8 cm (7 7/8 × 5 7/16 in.). The Cleveland Museum of Art, Charles W. Harkness Endowment Fund 1930.568 © Courtesy of Plattsburgh State Art Museum/Estate of Sally Kent Gorton.

Rockwell Kent, *Hail and Farewell*, 1930, wood engraving, platemark: 20.3 × 14 cm (8 × 5 ½ in.). The Cleveland Museum of Art, Charles W. Harkness Endowment Fund 1930.568 © Courtesy of Plattsburgh State Art Museum/Estate of Sally Kent Gorton.

the tall rope ladders, wooden masts, and bowsprits belonging to a large wooden sailboat. The American Car and Foundry yachts they advertised, in contrast, were powered by gasoline-fueled motor and the flick of a hand on an easy steering wheel—not by wind, sail, and muscle. Like the *Moby-Dick* illustrations, the American Car and Foundry engravings are characterized by the tightly packed parallel white lines that indicate shade, light, and three-dimensionality. The old-fashioned wood engravings created a classic, traditional vernacular for the industrial company, which needed to adjust its mechanistic, modern brand image when it entered the yacht market. Kent's images facilitated these motor yachtsmen's adjustment to a system of enervating fossil modernity, while indulging their fantasies of nineteenth-century maritime masculinity in the muscled bodies of working sailors and the figure of the artisan craftsman that is invoked in the woodblock style.

These engravings speak precisely to the Arts and Crafts legacy as cultural historian T. J. Jackson Lears defined it: the forced migration of meaningful work to the realm of old-fashioned hobbies. The Arts and Crafts movement, while it revived certain craft practices, never achieved the structural transformation it promised. Instead, according to Lears, the movement simply eased alienated workers' incorporation into the bureaucratic workplace by enriching workers' leisure time rather than reforming their experience of labor.[75] Modern capitalism splits work and life, dividing work for pleasure and work for pay, forcing even would-be craftspeople to accept that authentic labor can be found only in unpaid recreation. The customers who would buy American Car and Foundry yachts were privileged yachtsmen who wanted to enrich their leisure time and transmit their wealth. Kent's nostalgic wood engravings evoked for American Car and Foundry what his woodblock colophons evoked for the publishing houses: a visual style that communicated taste and prestige rooted in cultural authority.

The American Car and Foundry engravings were no more the product of artisan labor than the modern motor yachts they advertised. Twentieth-century wood engravers did not simply take the wooden block they just engraved, set that block on the bed of a nearby press, and begin printing. That quaint, unalienated process did not exist in the world of commercial publishing in the early twentieth century, if it ever existed at all. Even in the context of new publishing technologies, several steps of mediation between the artist and the published book or magazine were required. Unacknowledged workers and artists other than Kent participated in the process of turning his engraved block into multiple prints on paper. Kent's was not the only hand to engrave the original wooden block, either:

he enlisted other artists, including Ione Robinson, to do the most labor-intensive engraving when he was pressed for deadline.[76] Kent almost never engraved a wooden block and printed directly from that block: "Practically all of my prints have been printed from electrotypes of the blocks. Between prints made from an electrotype and prints made from the block itself I have been able to discover no difference, even under the magnifying glass. Using an electrotype insures the original block against such damage to its fine lines as may easily occur despite the vigilance of the printer, and against the cracking of the block under the pressure that must be applied. Such cracking is a not infrequent occurrence."[77] The passive voice—"my prints *have been printed*"—obscures the labor of the workers who deployed machinery to translate Kent's blocks into electrotypes. And there is a defensive quality to Kent's explanation of how virtually identical the electrotype is to the "original" wooden block. Like the motor yacht's modern engine, wood engraving in the early twentieth century was the result of an industrial process and divided labor. The artisanship and autonomy that wood engraving evoked were nostalgic fantasies of the past. Kent's wood engravings reflect the unintended complicity of Arts and Crafts ideology with the division of labor in fossil capitalism.

Although Kent modeled himself on Monhegan's working fishermen, Kent himself shared more than he would have liked to admit with a wealthy leisure class that enjoyed motor yachting—with those customers of the American Car and Foundry to whom his advertisements called. Kent's experience of Maine, for example, was made possible not only by the visible vestiges of an older, better life but also by the tourist industry, which commercialized leisure. The paradox of Maine's privileged place in the work of such modernists as Hartley, Kent, and others is that Maine was only accessible on account of the development of a highly modern tourist infrastructure—a network of highways, gas stations, motels, and restaurants that was rapidly converting the land of authentic labor into hypermodern "Vacationland" (a slogan that first appeared on Maine license plates in 1936).[78]

Moby-Dick, AGAIN

Rockwell Kent restored the new, modern *Moby-Dick* of the Melville Revival to the wooden world of the nineteenth-century sea. His illustrations bestowed *Moby-Dick* and other newly classified classics of US literature a woodblock style that hearkened back to the Golden Day of US literature and of the wooden world of a bygone organic energy regime. In his own life, Kent drew as near as he could to the wooden world of nineteenth-century sailing ships. But there is a sharp split between the form and method of Kent's skeuomorphic illustration and between the fisherman's world of Maine he idealized and the fossil-modern world to which he sold his images of it.

Melville's *Moby-Dick* is a fractured text, too. In order to forecast the ends of the whaling industry, Melville made Ishmael nostalgic. Ishmael decided to forego the booming whaling port of New Bedford for the old whaling port of Nantucket and the huge whaling ships of the modern fleet for the "old fashioned claw-footed" *Pequod*. But there is a signal difference between Kent's nostalgia and Ishmael's. What is so surprising about Ishmael's nostalgia is that it is perverse, premature, out of time. And importantly, Ishmael's experience does not mirror Herman Melville's, for example. When Melville went whaling in 1842, he shipped out of New Bedford on a brand-new, state-of-the-art whaling vessel. Ishmael, by contrast, makes a concerted effort and spends dear money to leave New Bedford and ship out of Nantucket, an empty, sandy island populated by a strange community of old-fashioned Quakers. Nantucket becomes a space for Ishmael to reflect on energy transition: on what happens to boomtowns after the boom. Nantucket is a place out of time, a place where Ishmael can imagine the future, not the past. In Nantucket, Ishmael foresees the demise of the whaling industry from within the height of its productivity; he does not believe in endless economic growth and growing prosperity. He knows that Carthage was built by the survivors of Tyre's fall and that Rome would supersede Carthage. In 1851, *Moby-Dick* anticipates the obsolescence of American whaling and serves as a premature shrine to an industry that had not yet disappeared. Ishmael sees beyond the whaling boom to the end of the industry. He sees Nantucket, New Bedford, and the booming whaling industry itself as transient episodes in a panorama of rising and—always—declining empires. Ishmael's nostalgia led him to choose a ship slated for doom; the novel, of course, ends when the white whale sinks the claw-footed *Pequod*.

The splits and ruptures of fossil modernity itself come into view in Kent's *Moby-Dick* because Kent's illustrations and Melville's text engage in quite different politics of nostalgia, even while both valorize maritime labor. In Melville's presentation of *Moby-Dick*, nostalgia operates as a radical critique of capitalism, but in 1930, the skeuomorphic quality of Kent's illustrations mark his own accommodation to, and benefit from, fossil modernity. Kent's *Moby-Dick* is perhaps the ultimate expression of fossil modernity: drawn toward and built on the bones of a bygone organic energy regime and constructed completely from the technological, financial, and political structures of fossil capitalism.

Nowadays, *Moby-Dick* is still central to any discussion of labor and leisure in American culture, especially in light of recent literary scholarship about the importance of nineteenth-century maritime narratives for theorizing labor. Hester Blum, Margaret Cohen, and Amy Parsons have argued that narratives written by sailors in the nineteenth century—including, of course, Melville's *Moby-Dick*—offer a rich record and a unique corpus of thought about labor and the relationship of labor to other parts of human experience.[79] Here I extend that conversation by illuminating the material basis of labor and leisure practices to the energy resource regimes that constituted them: from wind, wood, muscle, and organic energy to fossil fuels, industrialization, and mechanization. Scholarship about the maritime world (including criticism of *Moby-Dick*) has always been implicitly recognized as "environmental" or "ecocritical": oceanic studies or the blue humanities are a recognized subdiscipline of environmental humanities scholarship. Kent's *Moby-Dick* helps establish the case for reading even those anachronistic, romanticized twentieth-century renderings of the nineteenth-century world as vital sites of environmental imagination—because in their nostalgia they represent the labor conditions of the present and the energy regimes that made them possible. And Kent's *Moby-Dick*, too, makes clear why labor itself is often inherently environmental, why meditations on how and why we do our work are also, at the same time, about the energy systems that our work require and sustain.

Kent's woodcut-style illustrations accept the invitation to nostalgia inherent in *Moby-Dick*, but they do not follow the novel's invitation to critique. Rather than untangling the energy-work nexus in fossil modernity, Kent's illustrations reveal just how tight that knot has been tied. Kent evoked his fantasy of artisan labor in the woodblock style of his illustrations, but the skeuomorphic drawings—made without a woodblock—acknowledge and even abet obsolescence itself.

EPILOGUE

The Bone in Our Teeth

From 1841 until its last voyage in 1921, the ship *Charles W. Morgan* was a working whaling vessel.[1] After its final whaling voyage, the *Morgan* languished in New Bedford harbor for five years.[2] In 1925, as described in chapter 4, the *Morgan* was moved from New Bedford to nearby Round Hill, Massachusetts, entombed in sand, and opened to visitors as a tourist attraction at the estate of millionaire Edward H. R. Green. When Green died in 1935, he left the *Morgan* unaccounted for in his will, and the corporation called Whaling Enshrined that owned the ship found itself without funds to maintain the old vessel. A hurricane in 1938 knocked the *Morgan* from its sandy bed at Round Hill and brought the question of the ship's preservation to a crisis. Maritime preservationist Carl Cutler offered to take on the ship and provide for its upkeep so long as the New Bedford corporation would relinquish it to the newly founded Mystic Seaport Museum in Connecticut. The *Morgan* was shored up for its voyage and towed south by a US Coast Guard vessel. On November 8, 1941, the ship arrived in Mystic. Since then, the *Morgan* has been the crown jewel of the museum's collection: the last extant whaling vessel from the nineteenth-century US fleet and, with its sturdy black hull and tall wooden masts, an undeniably majestic feature on the skyline of the small Connecticut town.

The *Charles W. Morgan* was a working whaling ship for thirty-seven voyages and seventy-nine years, but the ship has been something else—a monument, a movie set, a relic, a ruin, a museum—for even longer. In its transformation from extractive industry in the nineteenth century to service and tourism in the twentieth and twenty-first centuries, the ship encapsulates the region's economic history. In the eyes of some nostalgic visitors, the ship with its curved wooden hull evokes a bygone wooden world, an expansive Age of Sail, or an Arts and Crafts fantasy of hard, honest labor in nature. It would be easier for the *Morgan* to bear the weight of these fantasies if it had always been a leisure vessel with a gentler history, like the charter sailboats and expensive yachts that share its waters on the Mystic River and in Long Island Sound. But as a whaler, the *Morgan* also testifies to atrocity, the systematic slaughter of whales, exploited labor, the violence of US settler colonialism, and the whale oil industry that augured fossil modernity and climate crisis. As the poet Robert Lowell wrote about another beautiful and painful monument, the *Morgan* sticks like a bone in the museum's throat.[3]

In order simply to float, to fulfill their most basic state of being, ships require constant maintenance and repair. Most ships, when they run out their usefulness, are salvaged: dismantled, their parts put to new use. Even those ships that are wrecked and sink to the bottom of the ocean floor are salvaged, their parts and soggy cargo metabolized through markets and put to new purposes. Media theorist Steven Jackson considers the invisible work of shipbreaking in an essay that proposes "broken world thinking" or "tak[ing] erosion, breakdown, and decay, rather than novelty, growth, and progress, as our starting points in thinking through the nature, use, and effects of information technology and new media."[4] Jackson meditates on the epic photographs taken by Edward Burtynsky of shipbreakers in Bangladesh as a model for broken world thinking, an example of how to make visible the work of dismantling and undoing. Jackson reminds readers that this is what happens to almost all ships: "Save for the special cases of hostile sinking, shipwreck, or honorable retirement and preservation, it was this: they were disassembled, repurposed, stripped, and turned into other things, in sites and locations like the shipbreaking beaches of Bangladesh that have dropped out of history and imagination."[5]

Unless its status as a "special case" is perpetually renewed, the *Charles W. Morgan* will follow all other ships to a fate in wreck, ruin, or salvage. Owing to the absolutely ordinary processes of weather and wear that break down all vessels, the *Morgan* found itself in 2009 in need of repair. The museum began fundraising for the ship's restoration in the wake of the 2008–9 recession that had left the museum's finances and future shaky. Rather than scaling back restoration plans to suit the museum's tight budget, the museum under the leadership of new president Stephen C. White decided to raise the stakes. Rather than maintain the ship only to let it float at dock, they decided to sail the ship on its "38th Voyage" after it underwent radical restoration. Sailing the *Morgan* to wider audiences on the East Coast would be a bold gambit to reenergize the museum and recapture the public's attention. White called the *Morgan*'s restoration and 38th Voyage a "moonshot." I have since asked White what he meant when he called the voyage a "moonshot," and he explained that he was a "child of the 60s" and a "Kennedy fan," and he thought that the idea of sailing the *Morgan* was "a big idea that would lift us up."[6] The *Morgan* needed to be restored, and the cash-strapped museum needed to inspire fundraisers for the enormous job. To *sail* the *Morgan* was a more inspiring, galvanizing idea than simply to restore it for a sedentary life at dock. The *Morgan*'s fate depended on the 38th Voyage; the ship needed the voyage in order to inspire fundraisers to support its restoration, and the ship needed the restoration in order simply to continue existing. The 38th Voyage was an existential risk for the ship, and possibly even for the museum.

In the summer of 2014, the *Charles W. Morgan* whaling ship was towed out of its museum berth, past the raised drawbridge in the center of Mystic, down the widening Mystic River and into Long Island Sound. There, eventually, the *Morgan* dropped its tow line and sailed for the first time in almost a century. During sea trials before the 38th Voyage, the ship's crew found that the *Morgan* sailed much faster and handled with more agility than anyone had ever imagined it could. In the summer of 2014, the *Morgan* sailed along the East Coast, calling at New London, Connecticut, Newport, Rhode Island, and, in Massachusetts, Martha's Vineyard, New Bedford, Provincetown, and Boston, where it docked alongside the S. S. *Constitution*, "Old Ironsides." On August 6, after two and a half months of voyaging, the *Morgan* returned to its dock at the Mystic Seaport Museum, and there it floats today.

I sailed with the *Morgan* in 2014. The staff at Mystic Seaport selected me as a "38th Voyager," one of a cohort of passengers who applied for the honor of joining a short leg of the voyage. Among the 38th Voyagers were scholars, artists, scientists, and even Herman Melville's great-great-grandson. Curators and historians in Mystic (as well as in New Bedford and Nantucket) had long worked to dispel the early twentieth-century whitewashed myth of Yankee whaling, and to bring visibility to the Indigenous, Portuguese, Black, and immigrant histories of New England whaling. The 38th Voyage reflected their efforts at telling a more accurate history of diversity among the whaling labor force; among the 38th Voyagers were descendants of the Indigenous, Portuguese, Black, and white crew members who lived and labored on the ship.

In 2014, I was developing this book project, and I already knew that I wanted to tell the story of the transformation of the *Morgan* from a working whaler to a museum ship. In my application for the 38th Voyager program, I wrote—perversely, considering that my knowledge of whaling history and culture was my sole credential for the opportunity—that the *Morgan*'s voyage would not teach me anything about the past but that it would offer a special opportunity to reflect on energy in the present. I posed the question in my application: "How do we reconcile our desire for energy with the environmental destruction our energy industries wreak?"

My few hours aboard the *Morgan*, I wrote, would help me answer that question. I was as naive as the wide-eyed greenhand sailors from Vermont and New Hampshire whom Ishmael passed on the streets of New Bedford at the beginning of *Moby-Dick*: "In some things," says Ishmael, "you would think them but a few hours old."[7] Since I wrote that earnest application, Hurricane Maria wrecked Puerto Rico, and Paradise, California, burned to the ground. Hurricane Sandy drowned New York, and Tropical Storm Allison flooded Houston. The Covid-19 pandemic slipped into the grooves cut by environmental racism, devastating exactly those communities of Black people, immigrants, and the poor whose lungs were already weakened by air pollution. As I write these words, the sky over my midwestern town is hazed with smoke from unprecedented wildfires in the Pacific Northwest. It is only an imagined, privileged "we" who desire more energy than ever, seemingly heedless of the galloping violence of fossil-fueled environmental destruction. The *Morgan*'s 38th Voyage did not bring—could not have brought—reconciliation; it raised, like ghosts, old and new stories about energy.

I met the *Morgan* in Newport, Rhode Island, the night before I would sail with the ship from Newport to Martha's Vineyard. It was early evening and foggy when I met my fellow 38th Voyagers for a picnic dinner on the dock. I was astonished by my shipmates' generosity; one 38th Voyager brought a stock pot filled with seafood for everyone to share, and another had fabricated whale-shaped souvenirs for everyone he met. It was the first and last day of camp all at the same time: we introduced ourselves only moments before promising to keep in touch. When I boarded the ship, fog condensing on the rigging dripped like rain on the wooden deck. I chatted with Voyagers and crew members who milled about on deck. One young crew member told me he had sailed with the *Morgan* since the ship's sea trials in late May and early June. He said it was miraculous to see the ship sailing with "the bone in her teeth." I did not know what the old-fashioned phrase meant, so he taught me. A ship sailing with the bone in her teeth is one that is sailing quickly, kicking up a wave of white foam at the front of the ship that is supposed to look like a bone in an animal's mouth. Since then, I have encountered the term in nineteenth-century maritime narratives: Melville in *Benito Cereno* personifies one small boat as a dog he calls *Rover*, sailing with "a white bone in her mouth."[8] There were a lot of moments like this on the *Morgan*: people shy but teacherly, speaking old nautical jargon into new life.

It was a signal privilege of the 38th Voyager to spend the night before the voyage in the ship's forecastle—the large communal bunk room below decks and "before the mast" where ordinary sailors lived and slept. I chose a lower bunk tucked in a corner by the wall and then immediately changed my mind for a less claustrophobic upper bunk. Before I turned in for the night, I found the charging station outside the forecastle and added my iPhone to a pile of electronic devices sucking power through the wires tacked onto nineteenth-century lumber. I was afraid I'd be too excited to sleep, but I did, having exhausted myself like an overexcited toddler. When I awoke in the morning, the ship prism in the deck above my head was shining bright white, and I could hear the crew outside preparing breakfast. I took my phone off the charging wall and, as I stood there below deck, my body under the water line in the belly of a whaling ship, I felt closer than usual to the creatures whose fossilized remains, turned into coal and gas and oil, had fed my phone overnight. In the gloom belowdecks, my iPhone glowed like a bone.

The *Morgan* left Newport Harbor under tow, attached by a heavy hawser line to the tugboat called the *Sirius*. As soon as we left the dock, I felt the weight of the 38th Voyagers' responsibility to *experience the voyage*. I jumped into a line of crew members to help haul a heavy piece of rope, but I did not fall into the right rhythm of labor; I was in the way. So I turned to the work I knew how to do: interviewing my fellow Voyagers and crew members about the *Morgan*'s voyage, and what it meant to them. Within a few minutes, seasickness stopped me in the middle of my conversation. The wind was light, and my lungs were filled with the diesel exhaust from the *Sirius*. People give you a lot of advice when you're seasick, and I took it. I drank weak tea, ginger ale, and lots of water. I ate ginger candy, half a sleeve of saltine crackers, and a handful of a fellow Voyager's Dramamine pills. One shipmate told me that all I needed to do was "go with the motion"—which was easy for him, he said, because he was a "Zen bastard." I told him that I was a sick bastard.

Finally, the *Morgan* dropped its tow line, separated from the *Sirius*, and began to sail under its own power. No hawser line, no diesel exhaust, no engine drone, and all at once, no seasickness. It was as if the ship's muscles suddenly engaged. All I could hear was the slap of sails in the wind and the shouts of crew members. I rejoined the frenzied sociality of the Voyagers, comparing notes, conducting breathless interviews. I felt a sense of awe for the strong crew members who operated the massive ship and for the labor of the curators and staff who made this logistical fantasia a reality. As I had predicted, I did not feel particularly close to the laborers of the past. After all, that morning I stood at a bathroom sink next to one of those strong crew members now hauling heavy lines. We had wished each other good morning; she washed her face with a cleanser from Trader Joe's.

Even under that day's light wind, the ship sailed faster and faster. We drew closer to Martha's Vineyard, and the *Morgan* picked up an entourage of private boats, from dinghies to yachts, that chased the huge whaling ship into harbor. People on shore blew air horns and vuvuzelas, fired canons, and spouted water, and even though the ship was still moving carefully into the harbor, the voyage was already over.

I stayed on Martha's Vineyard for a few days after my voyage. At each port of call on the Morgan's 38th Voyage, Mystic Seaport staged a dockside exhibition to educate visitors about the *Morgan*, the history of whaling, and the current environmental health of the ocean. The exhibition was staged in a grouping of small tents. One tent held a small stage where musicians played sea chanteys; another tent housed scientists and marine archaeologists from NOAA (National Oceanic and Atmospheric Administration) who provided information about the state of the world's oceans; in yet another tent, a video about the *Morgan*'s history played on a continuous loop. There was a wooden whaleboat that visitors could climb aboard and, on the other side of the exhibition (too far away for harpooning selfies), a life-size inflatable whale. One tent housed an interactive exhibit called "What Bubbles Up?" where visitors wrote reflections about whales and pinned them onto the wire-mesh whale. When I visited the dockside exhibition in Martha's Vineyard, most of the messages seemed to have been written by kids with environmentalist anxieties: "whale meat is horrible"; "stop polluting ocean!"; "stop hunting for whales." Others expressed the sunbaked thoughts of bored vacationers: "was up?" and "keep it up whale!" Mystic Seaport curators worked with local institutions in each port to adapt the dockside exhibition to local histories and concerns. On Martha's Vineyard, for example, members of the Wampanoag Tribe of Gay Head (Aquinnah) staffed a tent to educate visitors about the nation's abiding relationship with whales and whaling, from whale harvesting on the island long before colonial settlement to the deep involvement of Wampanoag families in the commercial industry of the nineteenth century. As I rode the bus back and forth between the *Morgan* at dock and the hostel on the far end of the island, I heard fellow visitors talking about the ship; most of them called it the *Captain Morgan*, like the rum.

Several days later, I rejoined the *Morgan*'s 38th Voyage at New Bedford. The *Morgan*'s installation in New Bedford was different than it had been on Martha's Vineyard; the NOAA people, the chanteymen, the video, and the inflatable whale were all there, but the exhibition was set up on a concrete commercial pier alongside the hundreds of working vessels in New Bedford's fishing fleet. On June 28, the city staged a civic celebration called the New Bedford Homecoming in honor of the *Morgan*'s return to its home port. On that day, the ship and the dockside exhibition were flooded, not with island vacationers in baseball caps, but with local families, local politicians visiting with constituents, and big groups of children in traditional Azorean and Cape Verdean clothes queuing up for the parade.

New Bedford's economy has not always thrived in the fossil modernity of the twentieth and twenty-first centuries. The textile manufacturing industry that succeeded whaling as the city's main industry in the late nineteenth and twentieth centuries did not enrich the city the way whaling had once done. Unlike Nantucket, which galvanized its own postindustrial quaintness to become one of the most exclusive and expensive tourist resorts in the country, New Bedford did not develop a tourist industry, apart from a few blocks in the historic downtown that host the New Bedford Whaling Museum and the Whaling National Historical Park. The city is home to one of the largest commercial fishing fleets in the nation—continuing New Bedford's history in extractive industry—yet fishing is embattled, too. But the city is ever hopeful for a renaissance, and the civic leaders of twenty-first-century New Bedford are pinning their hopes on another energy industry from the sea: offshore wind.

The *Morgan*'s homecoming to New Bedford gave civic leaders an opportunity to reflect on the history and prospects of energy in the region. Massachusetts senator Elizabeth Warren gave some of the first remarks at the ceremony:

> The reason that we think about the *Charles W. Morgan* is that it was the *Morgan* and the whaling industry that made New Bedford "the city that lit the world." Now the *Morgan* returns, and the city that lit the world will do it once again. That's what New Bedford will do. But this time when New Bedford lights the world instead of whaling oil it will be with the lights from wind energy. That's what New Bedford will do. . . . This abundant and renewable resource will provide clean energy for all of America. It will provide hundreds of jobs. It will lower our dependence on fossil fuels and it will permit New Bedford to expand its shipping capacity and create more economic growth right here in New Bedford. This is about the past but today is about the future as well. The sea has always been an economic engine for New Bedford and it will continue to be so. Through these investments in infrastructure, in wind energy and support for our local fishermen, New Bedford will be the future.[9]

A prosperous New Bedford powered by offshore wind energy was, in 2014 when Warren spoke those words on the pier, a future hope. Construction was still underway on the Marine Commerce Terminal, a new heavy-lift cargo facility in New Bedford Harbor that would serve as the staging area for the emerging wind industry. At that point, no wind energy companies had yet secured a permit to build wind turbines or produce power, and so the prospects for the terminal were uncertain. The construction of the terminal was completed in 2015, one year after the *Morgan*'s visit. Wind energy faces formidable challenges. Republican donors and oil and gas companies are spreading disinformation about climate change and funding expensive lawsuits against offshore wind developers in order to protect the fossil fuel industry.[10] Locally, fishermen fear that wind infrastructure will disrupt fishing operations and further endanger the New Bedford economy, while wealthy NIMBY landowners on the coast and islands oppose wind turbines that might mar sea views. You could hear Elizabeth Warren speaking to these constituencies during her remarks at the *Morgan* in 2014: to wind energy promoters, she promised that the work will come soon; to fishermen, she promised that the government will not forget them; and to New England vacationers, she drew a line from offshore wind power to the New England history and heritage embodied in the beautiful lines of the *Morgan*.

New Bedford mayor Jon Mitchell spoke in even more personal terms, characterizing the *Morgan* as an emblem of the city's history and its return as a harbinger of brighter futures.

> The *Morgan* is much more than a tourist attraction, of course. We've all heard it said over the years as we've discussed the fortunes of this city; people have commonly said: "Well, if it weren't for the loss of the *Morgan*. . . . Well, if we'd only kept the *Morgan*." And so it has been seen as this emblem of decline over the years, of our economic struggles over many decades. Now, that's somewhat overstated of course; we have lost industry in our city, in our region, for reasons having much more to do with our response to foreign competition than a whaleship that can't go whaling again.
>
> But this type of collective memory means a lot to the community. So this is the significance of the occasion. Nostalgia's a very powerful thing. We look back to the past with some pain, and the word "nostalgia" literally translated from the Latin means "our pain." But New Bedford nowadays is a city that's looking forward. It's looking forward to a much brighter future. Once a city that was trailing behind the pack, we are now leading the pack. This is

a city that has been recently depicted in national newspapers and magazine as tops in the nation when it comes to renewable energy. We are of course the largest commercial fishing port in America, we are about to become the birthplace of offshore wind. . . . The *Morgan*, the return of the *Morgan* right now is a moment for us to turn the page. This is where our past meets our great future. The *Morgan* reminds us more than anything else right now of the difference between living in the past and drawing inspiration from our history.[11]

It is telling that Mitchell associated the return of the *Morgan* to New Bedford as an invitation for the city to confront the pain in its past. The whales might have scoffed at the mayor's earnest sentimentalism about the *Morgan*, but Mitchell that day joined long tradition of New Englanders who fashioned whaling history into a language for talking about the fortunes, the identity, and even the feelings associated with changing energy regimes. The nostalgia Mitchell talked about is akin to the extractivist nostalgia I have discussed earlier; the period of past prosperity Mitchell is nostalgic for was predicated on the extractive whaling industry.[12] But extractivist nostalgia is different on the other side of the twentieth century, in the city that fossil modernity has thoroughly exhausted.

Wind energy offshore southern Massachusetts is still a hope, albeit one slightly less distant than it seemed in 2014. In May 2021, the Biden administration gave final approval for the construction of the offshore wind farm to a company called Vineyard Wind.[13] Since then, wind energy infrastructure projects have been breaking out all over southern New England. Vineyard Wind leased New Bedford's Marine Commerce Terminal in 2021.[14] In 2023, the underwater cables that will deliver electricity from offshore windmills were laid on a Cape Cod beach where all that power will eventually come ashore.[15] A New Bedford waterfront site has been sold to developers who plan to build another marine terminal for offshore wind staging. The Foss Marine Terminal is slated to be built on the site of the old Eversource Energy/Sprague Oil property, which had been the site of a manufactured gas plant and a tar processing facility. A small crowd of spectators, including Mayor Mitchell, cheered when the plant's emblematic smokestack came down in January of 2023.[16] But this is all still speculative infrastructure for an energy industry that does not quite yet exist. New Bedford does not yet know if it can "turn the page" and welcome an energy regime, a new source of economic prosperity, and step away from the violence of fossil fuels and the pain of the last century.

What will happen to the remnants of our current energy regime when—if—we fully relinquish fossil fuels, fossil capital, fossil modernity? Territorial Agency, an independent organization combining architecture, spatial analysis, and advocacy, proposes a Museum of Oil. "The Museum of Oil's aims are . . . rather solid: to put the oil industry into the museum. To make it a thing of the past."[17] Territorial Agency, headed by architects John Palmesino and Ann-Sofi Rönnskog, operates the Museum of Oil as an ongoing project but not as a permanent place open to visitors. The Museum of Oil pops up only on occasion for temporary exhibitions, as in 2019 for the Chicago Architecture Biennial. The exhibition in Chicago consisted of several huge panels grouped in a vast hall. The panels were cantilevered to tip forward, tilting and looming over visitors. Each monumental panel was printed with a huge aerial photograph or satellite image of an oil site: boreal forests in Alberta, the refinery town of Whiting, Indiana, the damaged zone around the Deepwater Horizon oil disaster. Digital screens propped up in front of the panels projected photographs, maps, schematics, and text that aimed to make clear the relationship between specific sites of extraction or production and big global oil infrastructure and systems. Territorial Agency produces its museum by analyzing Landsat images and data generated by orbiting multispectral scanners; it then visually elaborates those images in order to render the oil industry and its effects on the planet visible. These images, along with reams of data and images chronicling the effects of oil production, refining, and consumption, comprise the exhibits of the Museum of Oil, but, as with all museums, most of their work happens outside of visitors' view. The Museum of Oil makes the slow carnage of oil extraction visible, serving as a site of resistance against fossil capitalism and extractive violence. Palmesino and Rönnskog's project has stakes for the theory of the museum itself, too. They call their practice one of "radical conservation," even though their museum does not collect or display the material objects and structures it exhibits through images. "Museums aren't passive, if grand, high culture warehouses," write Palmesino and Rönnskog. "They don't merely provide a platform for our collective cultural goods—they locate power as well as artefacts."[18] A practice of radical conservation, the aim of which is to locate power, is a promising model for dealing with the remains of the fossil fuel industry.

But what indeed of those material remains—of mines, wells, refineries, pipelines, gas stations? Energy scholar and activist Jeffrey Insko was called to energy activism in 2010 when the Enbridge 6b pipeline that transports diluted bitumen from the Alberta oil sands ruptured near Marshall, Michigan, and spilled into a tributary of the Kalamazoo River.[19] The Enbridge 6b pipeline runs through Insko's backyard in Michigan, and when Enbridge applied for permits to build a newer, ostensibly safer pipeline, he began organizing a group of residents—and, eventually, regional and national climate change and Indigenous activists—to advocate for pipeline safety and oversight. Enbridge did in fact build that new pipeline through Insko's backyard, but their promise to "replace" the old one was euphemistic; there is still an old decommissioned pipeline, one nominally maintained by Enbridge, running alongside the new one. Insko asks: "How do we think about infrastructures that no longer perform their function?"[20] As for the very real, very substantial old pipeline running through his backyard, still maintained by a pipeline company, Insko calls for dismantling: "I'd prefer to see the damn thing removed altogether. Part of me, in fact, would rather it just rot in place."[21] Insko's call echoes that of artist Jenny Odell, who in the book *How to Do Nothing* (2019) proposed "manifest dismantling" as an antidote to the destructive work of development wrought by the extractivist settler ideology of Manifest Destiny. Odell's signal example of manifest dismantling is the removal in 2015 of the San Clemente Dam, like Insko's pipeline another infrastructure of fossil modernity.[22] Insko's proposal for dismantling the ruins of fossil modernity, and Territorial Agency's call for radical conservation offer provocative and urgent invitations for imagining the end of fossil fuels, perhaps even of hastening their obsolescence. Radical conservation and dismantling avoid the problems of preservation perhaps best encapsulated in one of the mottos of the Nantucket Preservation Trust, an organization that advocates for Nantucket homeowners to preserve historic interiors rather than gut-rehabbing them: "Gut Fish, Not Houses." Must something always be gutted in preservation? Is gutting the only option? Through dismantling and radical conservation, we might begin to deem the infrastructures of violence themselves not worth preserving and turn our conservation energy to the living world.

If nothing else, the history of whaling's cultural afterlife in fossil modernity reveals this: that old ways of being never disappear. It is not within the imaginative scope of extractive capitalism to understand its own limits, its own ends, or to deal with the hulking material structures it leaves behind in the abstract market turbulence of creative destruction. Even in the maritime world, where salvage is a time-honored process of rendering material remains back into systems of capitalist exchange and flow, old systems linger and give shape to our thoughts for the new. There may not be a single whaling ship or wharf left on Nantucket, but the island's whaling heritage is imprinted on every tourist dollar and real estate deal as surely as on the island's ubiquitous whale T-shirts. The history of whaling lingers in every "barrel" of oil, and even in Donald Trump's specious calls to save the coal industry and bring back "real America." And of course, whaling history lingers in the *Charles W. Morgan*, one of those "special case" ships that continues to be maintained in the service of some other use than that for which it was built.

The labor of maintaining an old piece of infrastructure like the *Morgan* is formidable; I can promise that whenever a reader encounters these words, someone in Mystic, Connecticut, will be caulking seams, scrubbing decks, or repainting the hull of the *Morgan*—or, at the very least, worrying about how to do it and how to pay for it. The *Morgan* has now served a purpose other than whaling for many more years than it served as a platform for extracting whale oil. History museums such as Mystic Seaport have seen declining admissions and financial trouble since their twentieth-century heyday. History, heritage, and tourism have their own unpredictable life cycles, too, and the *Morgan* perpetually faces a new obsolescence.

During my own short time on the *Morgan* during its 38th Voyage, I became fascinated with imagining where the *Morgan* might be had it not entered the hereafter of the museum. At the *Morgan*'s dockside exhibition Martha's Vineyard, I shared my perverse worry with marine archaeologist Matthew Lawrence from NOAA. What would the *Morgan* look like right now, on the ocean floor, if it had sunk like the *Wanderer*? His first instinct was to say that nothing would be left; he reminded me that even shipwrecks are salvaged. I pressed him further, and he said, "Think green." He said to imagine that everything left would be tinted green due to the verdant vegetable life that grows in the nutrient-rich waters off the Atlantic Coast, that any extant structure would be cloaked with "a short seaweedy covering." He said that the average diver would swim over it without even knowing it was there.[23] The image makes me sad; I confess I love the ship I have known for years, slept on and sailed in, and I loved the community of brave laborers and enthusiastic fellow amateurs it brought together for the 38th Voyage. I love the ship that the *Morgan* has become, the beautiful wooden vessel that invited me to critique history—not just celebrate it—and that urges me into questions and concerns. But the image of an indistinct seaweedy pile at the bottom of the ocean makes me hopeful, too, for the future of fossil fuels. As long as we are plugged into a circuit of fossil energy, burning the labor of our bodies and the fuel of our planet, we are racing along with the bone in our teeth. The time has come to put it down.

NOTES

Abbreviations
MSM Mystic Seaport Museum Collections Research Center, Mystic, CT
NBWM New Bedford Whaling Museum Research Library, New Bedford, MA
NHA Nantucket Historical Association, Nantucket, MA

Introduction

1. Morris, *Derrick and Drill*, 57–58.
2. "Coal Oil."
3. Jeremy Zallen offers the most comprehensive history of the illuminant marketplace of the early and mid-nineteenth century, of which whale oil was a part. Zallen, *American Lucifers*.
4. Quoted in Morris, *Derrick and Drill*, 42.
5. Hubbard, *Little Journeys*, 143.
6. Heath, "Whalers to Weavers," 13.
7. MSS 61: Akin family papers, NBWM.
8. Morris, *Derrick and Drill*, 70.
9. My account of the mid-nineteenth-century emergence of the concept of "energy" builds on the work of Cara Daggett in *The Birth of Energy*.
10. "We experience ourselves, as moderns and most especially as modern Americans, every day in oil, living within oil, breathing it, and registering it with our senses.... Oil itself is a medium that fundamentally supports all modern media forms concerned with what counts as culture." LeMenager, *Living Oil*, 6.
11. Timothy Mitchell has argued that the development of fossil fuels and democratic politics, for example, have been "interwoven from the start." Mitchell, *Carbon Democracy*, 8.
12. Wenzel, "Introduction," 7.
13. Zallen, *American Lucifers*.
14. Bonneuil and Fressoz, *Shock of the Anthropocene*, 101; Simpson and Szeman, "Impasse Time," 81.
15. "The consequences of these energy transitions went far beyond a simple set of substitutions in fuels. They created a shift from an organic to a mineral energy regime: a fundamental reorientation of the relationships between Americans and their environments that led to new ways of living, working, and playing." Jones, *Routes of Power*, 2–3; Tondre, "Conrad's Carbon Imaginary," 59.
16. "More than any other product, whale oil whetted the human appetite for clean, efficient, and affordable illumination." Black, *Petrolia*, 14.
17. Schooler, *Last Shot*.
18. Herman Melville eulogized some of those whalers taken over by the Union, the so-called Stone Fleet or Stone Blockade that was strategically scuttled in order to create a natural barrier in Charleston Harbor. Jones, "Navy's Stone Fleet."

> I have a feeling for those ships,
> Each worn and ancient one,
> With great bluff bows, and broad in the beam;
> Ay, it was unkindly done
> But so they serve the Obsolete—
> Even so, Stone Fleet!

19. Davis, Gallman, and Gleiter, *In Pursuit of Leviathan*; Bockstoce, *Whales, Ice, and Men*.

20. Davis, Gallman, and Gleiter, *In Pursuit of Leviathan*; Tønnessen and Johnsen, *History of Modern Whaling*.

21. The field of the energy humanities has a long history of ascribing or describing modernity as the product of intensive fossil fuel consumption. My term "fossil modernity" builds on Stephanie LeMenager's field-defining coinage "petromodernity," which describes "modern life based in the cheap energy systems made possible by oil." LeMenager, *Living Oil*, 67. I use "fossil modernity" instead of LeMenager's term to signal the complexity and interrelation of fossil fuel sources and the difficulty of isolating petroleum from oil or natural gas in narrating the historical changes wrought by fossil fuels in the late nineteenth and early twentieth centuries. LeMenager, *Living Oil*, 67. Imre Szeman and Dominic Boyer also claim that "to be modern is to depend on the capacities and abilities generated by energy." Szeman and Boyer, "Introduction," 1.

22. Tønnessen and Johnsen, *History of Modern Whaling*.

23. "Aboriginal Subsistence Whaling in the Arctic."

24. Demuth, *Floating Coast*.

25. On histories of Indigenous whaling within and beyond the commercial whaling industry, see Shoemaker, *Living with Whales*; and Shoemaker, *Native American Whalemen*.

26. "Modernity" has an especially fraught history in the scholarly field of energy history. Amitav Ghosh in *The Great Derangement* takes issue with how the Pennsylvania oil boom of the 1860s is so often cited in energy history as the beginning of the modern oil industry and points to the development of the centuries-old oil industry in Burma as no less "modern." Modernity, in Ghosh's view, is another term for Western exceptionalism: "The one feature of Western modernity that is truly distinctive [is] its enormous intellectual commitment to the promotion of its supposed singularity." Ghosh, *Great Derangement*, 103. With regard to Indigenous history in the North American context, I am particularly influenced here by Mark Rifkin's characterization of modernity as a feature of "settler time": "notions, narratives, and experiences of temporality that de facto normalize non-native presence, influence, and occupation." Rifkin, *Beyond Settler Time*, 9.

27. Wenzel, "Introduction," 7.

28. Williams, *Marxism and Literature*, 122. In my study of obsolescence, I am also indebted to the work of Francesco Orlando, although I do not take up his concept of the "nonfunctional" because I argue that obsolete cultures still perform work. Orlando, *Obsolete Objects*.

29. Acland, *Residual Media*.

30. Acland, *Residual Media*, xiii, xiv.

31. Acland, *Residual Media*, xv.

32. Schumpeter, *Capitalism, Socialism, and Democracy*; McCraw, *Prophet of Innovation*.

33. Shukin, *Animal Capital*.

34. Shukin, *Animal Capital*, 20–22.

35. Shukin, *Animal Capital*, 21.

36. On the illuminant market in the eighteenth and nineteenth centuries, see Zallen, *American Lucifers*.

37. Buell, "Energy Systems," 140.

38. Szeman and Boyer, *Energy Humanities*; Szeman, Wenzel, and Yaeger, "Whaling"; Yaeger, "Editor's Column"; LeMenager, *Living Oil*; Johnson, *Carbon Nation*; Jones, *Routes of Power*; Wenzel, "Petro-Magic-Realism" (2006); Szeman, Wenzel, and Yaeger, "Whaling"; Barrett and Worden, *Oil Culture*; Szeman, *After Oil*. An important work of energy humanities scholarship that brought whale oil culture into productive conversation with petroculture was the article by Heidi Scott: "Whale Oil Culture" (2014).

39. Daggett, *Birth of Energy*.

40. I use the terms "intensification" and "acceleration" advisedly and with reference to specific concepts that have arisen in energy historiography. I am thinking specifically of Christopher Jones's explanation of fossil fuel consumption as a process of "intensification," in which fossil fuel extraction and distribution became processes that themselves relied on fossil fuel power, creating "a set of positive feedback loops [that] arose between the building of infrastructure, the economic investments in these systems, the actions of human agents, and new consumption practices." Jones, *Routes of Power*, 8. "Acceleration" nods to the idea, current among promoters of the Anthropocene concept that "human impact on the Earth and the biosphere . . . has escalated" in recent human history. The onset of the Anthropocene and the Great Acceleration are subject to debate, with some advocating for a start date with European colonialism in the fifteenth century, others advocating for the invention of the steam engine in the late eighteenth century, and others still advocating with nuclear weapons testing and deployment around 1945. Informed by widespread critiques of the Anthropocene concept, I do not pick a side in this particular debate but seek instead to ground my understanding of "energy" in massive fossil fuel consumption in the nineteenth century.

41. Daggett, *Birth of Energy*, 3.

42. Daggett, *Birth of Energy*, 33–50.

43. This notion of energy often circulates in works that deal with forms of energy *not* derived from fossil fuels, especially solar energy. Demuth, *Floating Coast*; Underwood, *Work of the Sun*; Szeman, Wenzel, and Yaeger, "Whaling"; Shannon, "Tallow"; Haynes, "Animal"; Starosielski, "Beyond the Sun."

44. Demuth, *Floating Coast*, 3–6.

45. Demuth, *Floating Coast*, 308.

46. Tompkins, "Sweetness, Capacity, Energy"; Frost, *Biocultural Creatures*.

47. Zallen, *American Lucifers*. Scott, "Whale Oil Culture."

48. Simpson and Szeman, "Impasse Time," 80–81.

49. I am inspired by the energy futures imagined and discussed in the special issue on "Solarity" of the *South Atlantic Quarterly*: Szeman and Barney, "Introduction."

50. Jussi Parikka, Erkki Huhtamo, Shannon Mattern, Nicole Starosielski, Wendy Hui Kyong Chun, and others. Like these scholars, I build on the distinction between history and archaeology first set out by Michel Foucault. Parikka, *What Is Media Archaeology?*; Huhtamo and Parikka, *Media Archaeology*; Starosielski, *Undersea Network*; Parks and Starosielski, "Introduction"; Starosielski, Soderman, and Cheek, "Introduction."

51. Acland, *Residual Media*; Williams, *Marxism and Literature*.

52. Huhtamo, "Dismantling the Fairy Engine," 28.

53. Parikka, *Geology of Media*.

54. Johnson, *Mineral Rites*.

55. Starosielski, *Undersea Network*.

56. Lisa Parks and Nicole Starosielski argued that "our contemporary mediascapes would not exist without our current media infrastructures." John Durham Peters argued that "media are infrastructures that regulate traffic between nature and culture"; Darin Barney has argued that "transportation is communication, and its infrastructures are media." Parks and Starosielski, "Introduction," 1; Peters, "Infrastructuralism," 32; Barney, "Pipelines," 268.

57. LeMenager, *Living Oil*, 11.

58. Peters, *Marvelous Clouds*, 2.

59. Nixon, *Slow Violence*.

60. Whyte, "Indigenous Climate Change Studies"; Whyte, "Dakota Access Pipeline"; TallBear, "Badass Indigenous Women Caretake Relations"; Estes, *Our History Is the Future*; Todd, "Fish, Kin and Hope"; Rifkin, *Beyond Settler Time*; Rifkin, "Duration of the Land"; LaDuke and Cowen, "Beyond Wiindigo Infrastructure."

61. O'Brien, *Firsting and Lasting*.

62. Todd, "Fish, Kin and Hope."

63. Mark Rifkin defines "settler time" as "notions, narratives, and experiences of temporality that de facto normalize non-native presence, influence, and occupation." Rifkin, *Beyond Settler Time*, 9.

64. Szeman and Boyer, *Energy Humanities*; Szeman, Wenzel, and Yaeger, "Whaling"; Yaeger, "Editor's Column"; LeMenager, *Living Oil*; Johnson, *Carbon Nation*; Jones, *Routes of Power*; Wenzel, "Petro-Magic-Realism" (2006); Szeman, Wenzel, and Yaeger, "Whaling."

65. Jones, "Petromyopia."

66. Szeman, "System Failure"; Whalan, "Oil Was Trumps"; Yergin, "Ensuring Energy Security."

67. LeMenager, *Living Oil*.

68. Susan Leigh Star first articulated that infrastructure is invisible until it breaks down. That concept has been widely repudiated, but it is worth sustaining Star's effort to locate where, when, how, and to whom infrastructures are visible. Star, "Ethnography of Infrastructure."

69. McClintock, "Monster"; McClintock has also written about the Deepwater Horizon disaster in "Slow Violence and the BP Oil Crisis in the Gulf of Mexico."

70. Sekula, *Fish Story*, 74; Rozwadowski, *Vast Expanses*, 152–53; Madrigal, *Containers*.

71. Sekula, *Fish Story*, 51.

72. Rozwadowski, *Vast Expanses*, 152–53; Madrigal, *Containers*; Sekula, *Fish Story*.

73. Cohen, *Novel and the Sea*, 14.

74. Blum, "Prospect of Oceanic Studies," 670.

75. King, *Black Shoals*; King; Rozwadowski, *Vast Expanses*; King, *Ahab's Rolling Sea*; Starosielski, *Undersea Network*; Jue, *Wild Blue Media*.

76. Sharpe, *In the Wake*, 3; King, *Black Shoals*; Ibrahim, *Black Age*; Gumbs, *Undrowned*.

77. Sekula, *Fish Story*, 106.

78. Lauter, "Melville Climbs the Canon"; Spanos, *Errant Art of Moby-Dick*; Spark, *Hunting Captain Ahab*; Aronoff, "Cultures, Canons, and Cetology."

79. Insko, "On Dismantling."

Chapter 1

1. Melville and Parker, *Moby-Dick*, 57–58.

2. Borowy and Schmelzer, *History of the Future of Economic Growth*; Cohen and Todd, *Infinite Desire for Growth*; Schmelzer, *Hegemony of Growth*; Gordon, *Rise and Fall of American Growth*.

3. Schmelzer, Vetter, and Vansintjan, *Future Is Degrowth*; Macekura, *Mismeasure of Progress*; Neurath, *From Malthus to the Club of Rome and Back*; Meadows and Club of Rome, *Limits to Growth*; Malthus, *Essay on the Principle of Population*.

4. Endless growth seemed possible especially in the 1940s–70s, when the decreasing price and increasing supply of oil drove economic growth as measured by GDP. "Thus, for several decades, the supply of energy seemed unlimited and its costs, for many purposes, negligible. If the economy appeared capable of unlimited growth, this was for practical reasons the mid-twentieth-century energy regime." Mitchell, "Economentality," 498. See also Macekura, *Mismeasure of Progress*, 107–15.

5. Mitchell, "Economentality," 507.

6. Quoted in NPR Staff, "Greta Thunberg's Speech at the U. N. Climate Action Summit."

7. I elucidate this history in the Introduction. Cara Daggett argues that energy, as a concept that forges a certain form of labor and power together with the fossil fuels that make it possible, was born in the mid- and late-nineteenth century. Daggett, *Birth of Energy*.

8. Tønnessen and Johnsen, *History of Modern Whaling*.

9. This is true of some of the works of criticism to which this study is particularly indebted, perhaps because these are the works that grapple most sincerely with the question of *Moby-Dick*'s relevance to contemporary crises of environment and capitalism. Frederick Buell calls *Moby-Dick* "astonishingly prescient," and Cesare Casarino ascribes to Melville "premonitions of the future." Buell, *Writing for an Endangered World*, 205; Casarino, *Modernity at Sea*, 16.

10. Sterne, "Out with the Trash."

11. Melville and Parker, *Moby-Dick*, 39.

12. Wenzel, "Petro-Magic-Realism" (2006), 451; Coronil, *Magical State*.

13. Melville and Parker, *Moby-Dick*, 39.

14. Wenzel, "Petro-Magic-Realism" (2017), 495.

15. Wenzel, "Petro-Magic-Realism" (2017), 494.

16. Melville and Parker, *Moby-Dick*, 164.

17. Bob Johnson offers a Freudian reading of energy in late nineteenth- and early twentieth-century US culture. He writes that the violence of fossil fuel extraction is repressed in society's unconscious: buried socially, geographically, and psychically far from the centers of cultural production. Johnson, *Carbon Nation*, 163.

18. Melville and Parker, *Moby-Dick*, 270.

19. Melville and Parker, *Moby-Dick*, 39.

20. Melville and Parker, *Moby-Dick*, 222.

21. Jones, *Routes of Power*, 8.

22. Blum, "Prospect of Oceanic Studies," 670.

23. Star, "Ethnography of Infrastructure"; Larkin, "Politics and Poetics of Infrastructure"; Rubenstein, Robbins, and Beal, "Infrastructuralism"; LeMenager, "Infrastructure Again, and Always"; Anand, Gupta, and Appel, *Promise of Infrastructure*; Hurley, *Infrastructures of Apocalypse*.

24. Insko, "Extraction," 173.

25. LaDuke and Cowen, "Beyond Wiindigo Infrastructure," 253; Thomas Davis, John Levi Barnard, and Stephanie Foote also note the inherent violence in the ordinary functioning of infrastructure. Davis, "'Far from the Gulf Coast, but Near It, Too'"; Barnard and Foote, "Ruins of the Future."

26. Melville and Parker, *Moby-Dick*, 12.

27. On the history of London's water infrastructure, including the sixteenth-century London waterworks, see Bolton, *London Water Supply*; and Taylor and Trentmann, "Liquid Politics."

28. Melville and Parker, *Moby-Dick*, 281.

29. Melville and Parker, *Moby-Dick*, 279.

30. Melville and Parker, *Moby-Dick*, 279.

31. The scholar Miya Moriwaki also reads the "grand Erie Canal" passage as one in which Melville establishes the relationship between whaling and imperial infrastructures.

32. Samuel Otter, for example, reads the cetology chapters as Melville's effort to "[anatomize] the features of ethnological discourse" and to "[inhabit and defamiliarize] nineteenth-century corporeal inquiry." Otter, *Melville's Anatomies*, 133.

33. Jeremy Zallen surveys the many illuminant resources, including but not limited to whale oil, that were supplanted by petroleum products. Zallen, *American Lucifers*.

34. On oil colonialism in Nigeria, see the work of Ken Saro-Wiwa and the scholarship of Jennifer Wenzel. Wenzel, "Petro-Magic-Realism" (2017). On oil and North American settler colonialism, see Whyte, "Dakota Access Pipeline"; Whyte, "Indigenous Climate Change Studies"; Estes, *Our History Is the Future*; and Todd, "Fish, Kin and Hope."

35. Simpson, *Mohawk Interruptus*, 158.

36. LaDuke and Cowen, "Beyond Wiindigo Infrastructure."

37. Melville and Parker, *Moby-Dick*, 253.

38. Melville and Parker, *Moby-Dick*, 253.

39. *Stories about the Whale*, 2.

40. Estes, *Our History Is the Future*.

41. Estes, *Our History Is the Future*; Whyte, "Dakota Access Pipeline"; TallBear, "Badass Indigenous Women Caretake Relations"; Todd, "Fish, Kin and Hope."

42. Whyte, "Dakota Access Pipeline," 168.

43. Wilson, *Underground Reservation*; Mathews, *Osages*; Mathews, *Sundown*; Estes, *Our History Is the Future*.

44. Demuth, *Floating Coast*.

45. Melville and Parker, *Moby-Dick*, 64. Jean O'Brien explains how the myth of Indigenous extinction is an important strategy of settler colonialism in nineteenth-century New England. O'Brien, *Firsting and Lasting*; Dimock, *Empire for Liberty*, 115–19.

46. "Throughout the nineteenth century, the majority of native men living near the southern New England coast went on at least one whaling voyage in their lifetime. Many were career whalemen who spent twenty to thirty years at sea. They went on long voyages, usually leaving home for three or four years at a time, returning briefly only to go out again on another lengthy voyage." Shoemaker, *Native American Whalemen*, 14. See also the work by Jason Mancini to document the story of Indigenous whalers and mariners more generally in the Indian Mariners Project, https://indianmarinersproject.com/.

47. Shoemaker, *Native American Whalemen*, 24.

48. Christopher Freeburg writes that Melville is more explicit in his critique of settler colonialism and Indian killing in other texts—notably in *The Confidence-Man* (1857) and in his review of Francis Parkman's *The Oregon Trail* (1849). Freeburg, *Melville and the Idea of Blackness*, 26–43.

49. "Jorge Luis Borges once remarked that the absence of camels in the Koran reveals the book's authenticity. It has roots in a culture in which camels are taken for granted. By the same logic, the neglect of nature in contemporary Western social theory perhaps shows the extent to which the massive appropriation of natural resources upon which the modern world depends has come to be assumed as a fact of life." Coronil, *Magical State*, 21.

50. Demuth, "Turn and Live with Animals."

51. Herman Melville, passage from chapter 65, "The Whale as a Dish," quoted in Kane, "Citation in the Wake of Melville," 37.

52. Kane, "Citation in the Wake of Melville," 38.

53. Kane, "Citation in the Wake of Melville," 38.

54. "Such is the endlessness, yea, the intolerableness of all earthly effort." Melville and Parker, *Moby-Dick*, 57–58.

55. Barnard, "Cod and the Whale," 853.

56. Barnard, "Cod and the Whale," 864.

57. Melville and Parker, *Moby-Dick*, 339.

58. Michelle Neely attributes Ishmael's ability to dismiss the possibility of extinction to the "romantic idea of nature that defines the sea as spatially and temporally beyond humanity's ken." Neely, *Against Sustainability*, 108. See also Barnard, "Cod and the Whale," 864.

59. See especially chapter 6 in Burnett, *Sounding of the Whale*.

60. Miller, *Extraction Ecologies*, 7–8.

61. Miller, *Extraction Ecologies*, 9.

62. Miller, *Extraction Ecologies*, 8.

63. Allen, *Republic in Time*; Rudwick, *Worlds before Adam*.

64. Foster, "Melville and Geology"; Hillway, "Melville's Geological Knowledge"; Foster, "Another Note on Melville and Geology"; Barnard, "Cod and the Whale"; Jonik, *Herman Melville*.

65. Marveling in the vicissitudes of deep geological time—of the distant past and the potentially posthuman future—is also a feature of recent writing that explores the Anthropocene concept. Jones, "Oil: Viscous Time in the Anthropocene."

66. Rudwick, *Worlds before Adam*.

67. Morris, *Derrick and Drill*, 227.

68. Morris, *Derrick and Drill*, 249.

69. Denton, *Our Planet, Its Past and Future*, 313–14.

70. Melville and Parker, *Moby-Dick*, 337–38, 339.

71. Through a reading of *Moby-Dick* done through the lens of the short story "Sea Story" by A. S. Byatt, Michelle Neely locates other "spatial, temporal, and cultural errors that allow Ishmael to deny the possibility of extinction"—and she argues that those errors are "still with us." Neely, *Against Sustainability*, 111.

72. Hubbert, "Nuclear Energy and the Fossil Fuels."

73. Schneider-Mayerson, *Peak Oil*.

74. Schneider-Mayerson, *Peak Oil*.

75. Davis, Gallman, and Gleiter, *In Pursuit of Leviathan*, 43.

76. Burnett, *Sounding of the Whale*; Davis, Gallman, and Gleiter, *In Pursuit of Leviathan*; Tønnessen and Johnsen, *History of Modern Whaling*.

77. LeMenager, *Living Oil*, 3.

78. Melville and Parker, *Moby-Dick*, 337–38.

79. Melville and Parker, *Moby-Dick*, 99.

80. Melville and Parker, *Moby-Dick*, 107.

81. Schultz, "Melville's Environmental Vision in Moby-Dick," 109; Barnard, "Cod and the Whale," 865.

82. Melville and Parker, *Moby-Dick*, 52.

83. Todd, "Fish, Kin and Hope," 107.

84. Todd, "Fish, Kin and Hope," 107.

85. Melville and Parker, *Moby-Dick*, 21.

86. Melville and Parker, *Moby-Dick*, 68, emphasis added.

87. Melville and Parker, *Moby-Dick*, 68.

88. Wai-chee Dimock reads the mythical, heroic characterization of Ahab as a condemnation. She locates places in *Moby-Dick* where Ahab is characterized as a barbaric, othered figure, such as a "Khan of the plank," a "sultan," or a "Grand Turk": "More verdict than tribute, such allusions hardly describe Ahab: they merely brand him as a thing of the past. At once regal and barbaric, he takes his place among other candidates for extinction." Dimock's argument about the way that the novel condemns Ahab by relegating him to the past extends beyond Ahab to Nantucket whalemen in general, and even to Nantucket itself. Dimock, *Empire for Liberty*, 117–18.

89. I use "hipster" advisedly and with the twenty-first-century vinyl-hunting music lovers in mind. John Davis cites the obsolescence of vinyl, noting its "formal status of

technical incompatibility... but also a symbolic status." Ishmael's preference for Nantucket rests on a similar paradox. Davis, "Going Analog," 223.

90. Melville and Parker, *Moby-Dick*, 64.
91. Melville and Parker, *Moby-Dick*, 64.
92. O'Brien, *Firsting and Lasting*.
93. O'Brien, *Firsting and Lasting*.
94. Luther S. Mansfield and Howard P. Vincent write that the *Pequod*'s "worn appearance and age probably constituted the main differences between the *Pequod* and Melville's own whaleship, the *Acushnet*." Melville, *Moby-Dick*, 629.
95. For my interpretation of the relatively long lives of whaling ships, I am indebted to Mary K. Bercaw Edwards, Glenn Gordinier, and Quentin Snediker, director of the historical shipyard at the Mystic Seaport Museum. Whaling historians Davis, Gallman, and Gleiter offer a more fanciful interpretation of the durability of whaling ships: they write that "a twenty-five-year-old merchant ship was *really old*, while a thirty-year-old whaler—its hull and deck protected by ablutions of whale oil—was no more than *mature*." Davis, Gallman, and Gleiter, *In Pursuit of Leviathan*, 240.
96. Melville and Parker, *Moby-Dick*, 64.
97. It seems that a law of diminishing returns held in the second half of the nineteenth century, when the size of whaling ships began to edge up more slowly. By 1860, the whaling industry had already begun its severe decline, anyway, and whaling voyages were more likely to be launched on extant ships rather than expensive new vessels. Davis, Gallman, and Gleiter, *In Pursuit of Leviathan*, 225.
98. Sterne, "Out with the Trash," 21.
99. Melville and Parker, *Moby-Dick*, 93.
100. Miller, *Extraction Ecologies*, 28–29; Gómez-Barris, *Extractive Zone*.
101. Acland, *Residual Media*, xv.
102. Sterne, "Out with the Trash."
103. Casarino, *Modernity at Sea*, 64.
104. Fornoff, Kim, and Wiggin, "Coda."
105. Smithson and Flam, "Entropy and the New Monuments," 11.

Chapter 2

1. Jackson, "Rethinking Repair," 222.
2. Mss 1, New Bedford Cordage Company Records, New Bedford Whaling Museum, New Bedford, MA. www.whalingmuseum.org/collections/highlights/manuscripts/mss-1/.
3. Brown, *Inventing New England*, 110–17. Brown describes the active efforts of local developers to create Nantucket into a tourist destination.
4. Charlestown Mutual Fire Insurance Co., "Insurance Policy," MS236: Nantucket Lodgings, NHA.
5. There are important exceptions to this observation about energy humanities scholarship, including works by Stephanie LeMenager and Bob Johnson. LeMenager, *Living Oil*; Johnson, *Carbon Nation*; Johnson, *Mineral Rites*.
6. Marchand, Meffre, and Polidori, *Ruins of Detroit*; Austin and Doerr, *Lost Detroit*; Moore and Levine, *Detroit Disassembled*; Ewing and Grady, *Detropia*.

7. Leary, "Detroitism."

8. Anthony Fassi makes a similar observation about the availability of twenty-first-century industrial ruins to the aesthetic of the picturesque. Fassi, "Industrial Ruins, Urban Exploring, and the Postindustrial Picturesque."

9. Leary, "Detroitism."

10. Godfrey, *Island of Nantucket*, 156.

11. Aron, *Working at Play*, 49.

12. That the infrastructures of tourism were so visible to late nineteenth-century New England tourists counters an often-held truism of infrastructure, that it only "becomes visible upon breakdown." Star, "Ethnography of Infrastructure," 382.

13. Gassan, *Birth of American Tourism*, 6. Dean MacCannell also accounts for the production and circulation of texts and images tourist sites as one of the critical stages of "sight sacralization," the process by which a place becomes a tourist sight. Following Benjamin, MacCannell calls this the "mechanical reproduction" phase of sight sacralization: "It is the mechanical reproduction phase of sacralization that is most responsible for setting the tourist in motion on his journey to find the true object. And he is not disappointed. Alongside the copies of it, it has to be The Real Thing." MacCannell, "Sightseeing and Social Structure," 63.

14. Like many other emerging theorists of infrastructure, Christopher Jones asserts that "infrastructures, as historians of large technological systems have demonstrated, are social as well as technological. Cultural values—reflected in financial incentives, state regulations, moral sentiments, and ideas about what constitutes a good life—strongly influence which technologies get built and how they are used over time." Jones, *Routes of Power*, 8.

15. *Map of Nantucket: The Old Colony Line to Oak Bluffs, Martha's Vineyard, and Nantucket via New Bedford and Woods Holl*, MS397, NHA.

16. *The Popular Resorts and Fashionable Watering Places of Southeastern Massachusetts and Newport, R.I.*, MS397, NHA.

17. *The Old Colony; or, Pilgrim Land, Past and Present*, MS397, NHA.

18. *The Popular Resorts and Fashionable Watering Places of Southeastern Massachusetts and Newport, R.I.*, 3.

19. *The Popular Resorts and Fashionable Watering Places of Southeastern Massachusetts and Newport, R.I.*, 3.

20. *The Tip End of Yankee Land*, MS397, NHA.

21. Burleigh, *Old Colony Railroad*, 2, 6.

22. Burleigh, *Old Colony Railroad*, 10.

23. "Fall River Line Ticket Envelope," MS155, NHA.

24. Bernstein, *Racial Innocence*, 71.

25. "Steamboat wharf and the steamer CITY OF TAUNTON discharging passengers and cargo, circa 1890s," New Bedford Whaling Museum, Accession Number 1991.48.10.16.

26. F. H. N., "Letter from Martha's Vineyard."

27. *A Guide to Martha's Vineyard and Nantucket*, 18, MS397, NHA.

28. F. H. N., "Letter from Martha's Vineyard."

29. The word "lapidary" lurks in the word "dilapidated," which derives etymologically from the Latin meaning, "to scatter as if throwing stones." The stoniness inherent in the word "dilapidated" suggests an aesthetic of stillness and frozenness—an arrestment of

time—in ruins. "Dilapidate, *v.*," *OED Online*, Oxford University Press, December 2022, www.oed.com/view/Entry/52732.

30. Godfrey, *Island of Nantucket*, 13.

31. *The Sea Cliff: Island of Nantucket, Mass.*, pamphlet, MS397, NHA.

32. The quaint has some things in common with the picturesque, a notoriously broad category popularized by William Gilpin in eighteenth-century Britain. Quaintness is closely linked to what Carrie Tirado Bramen called the "urban picturesque," an aesthetic mode that took hold in representation especially of New York in the late nineteenth century: "The urban picturesque operated as a form of local color, which captured the Old World customs and peculiarities that existed in modernity." The urban picturesque and the quaint are of similar chronotopes, representing spaces that are paradoxically modern and old-fashioned. Bramen, "Urban Picturesque," 446. On the picturesque in the United States, see also Conron, *American Picturesque*; and Baker, "Hawthorne's Picturesque at Home and Abroad," 417–44.

33. Ngai, *Our Aesthetic Categories*, 1.

34. Ngai, *Our Aesthetic Categories*, 64.

35. Ngai, *Our Aesthetic Categories*, 59.

36. Harris, *Cute, Quaint, Hungry and Romantic*, 27–28.

37. Lavery, *Quaint, Exquisite*, xii.

38. Lavery, *Quaint, Exquisite*, 31.

39. Godfrey, *Island of Nantucket*, 330–31.

40. Drake, "Nantucket."

41. Northrup, *'Sconset Cottage Life*, 88.

42. Northrup, *'Sconset Cottage Life*, 92–93.

43. *Old Places and New People; or, Our Pilgrimage, and What We Saw*, 30, MS397, NHA.

44. Brown, *Inventing New England*, 118. Dona Brown traces two strands of Nantucket tourist-anthropology: those who sought in Nantucketers vestiges of a heroic, adventurous past, and those who mockingly observed the "quaint old salts."

45. *The Popular Resorts and Fashionable Watering Places of Southeastern Massachusetts and Newport, R.I.*, 33.

46. Ngai, *Our Aesthetic Categories*, 69.

47. Wyer, *Nantucket Characters*.

48. Johnson, *Carbon Nation*, xix.

49. Murison, *Politics of Anxiety*, 158–62; Murison, "'Paradise of Non-Experts,'" 36.

50. Beard, *American Nervousness*, iv.

51. Murison, *Politics of Anxiety*, 159. Justine Murison argues that there is a "crucial difference" between Beard's theory of nervousness and others that preceded it: that Beard conceived of the body as a "closed circuit," unlike the predominant theory of the "open body," open and vulnerable to environmental forces. Doing so was one way in which Beard differentiated himself and, by extension, the emerging professional field of neurology from spiritualists and mesmerists.

52. Murison, *Politics of Anxiety*, 159; Daggett, *Birth of Energy*, 7.

53. Murison, *Politics of Anxiety*, 79–85.

54. Beard, *American Nervousness*, 5.

55. Beard, *American Nervousness*, 99.

56. Berardi, "Exhaustion," 155.
57. Corbin, *Lure of the Sea*, 95.
58. Aron, *Working at Play*, 33.
59. Aron, *Working at Play*.
60. Beard, *American Nervousness*, 26.
61. Beard, *American Nervousness*, 26.
62. Beard, *American Nervousness*, 76.
63. *The Old Colony; or, Pilgrim Land, Past and Present*, 70.
64. Northrup, *'Sconset Cottage Life*, 45.
65. *The Old Colony; or, Pilgrim Land, Past and Present*, 66.
66. Brown, *Inventing New England*.
67. Cook, *Historical Notes of the Island of Nantucket*, 3–4.
68. Cohen, *Novel and the Sea*, 14.
69. Corbin, *Lure of the Sea*.
70. Philbrick, *James Fenimore Cooper*, 260.
71. Morison, *Maritime History of Massachusetts*, vii.
72. Sekula, *Fish Story*, 51, 12.
73. Sekula and Burch, "Notes for a Film."
74. Sekula, Khalili, and Buchloh, *Fish Story*.
75. Sekula, Khalili, and Buchloh, *Fish Story*, 12.
76. Godfrey, *Island of Nantucket*, 330–31.
77. Sekula, Khalili, and Buchloh, *Fish Story*, 64–65.
78. Sharpe, *In the Wake*, 26.
79. Sharpe, *In the Wake*, 29.
80. Miles, "Nantucket Doesn't Belong to the Preppies."

Chapter 3

1. Cronon, *Nature's Metropolis*, 26.
2. Jones, *Routes of Power*, 2–3.
3. Davis, Gallman, and Gleiter, *In Pursuit of Leviathan*, 42.
4. Davis, Gallman, and Gleiter, *In Pursuit of Leviathan*, 44. On the environmental and economic histories of the early Pennsylvania petroleum industry, see Brian Black, *Petrolia*.
5. Davis, Gallman, and Gleiter, *In Pursuit of Leviathan*, 44.
6. Jones, *Routes of Power*, 65–70.
7. Blum, *View from the Masthead*, 4. Blum and others follow the periodization of maritime culture and the "golden age" nomenclature established by Thomas Philbrick, whom Blum calls "the most comprehensive chronicler of American sea writing." Blum, *View from the Masthead*, 197. See also Berger, *Antebellum at Sea*; and Cohen, *Novel and the Sea*.
8. Bolter and Grusin, *Remediation*; Henning, "New Lamps for Old."
9. Parks and Starosielski, "Introduction," 5. See also Nicole Starosielski's important contribution to oceanic studies with her chronicle of undersea cables and fiber optic networks, *Undersea Network*. Anticipating the current boom in infrastructure studies, Helen Rozwadowski offered a history of the transatlantic telegraph cable in *Fathoming the Ocean*.
10. Smith, "Thou Uncracked Keel."

11. Jue, *Wild Blue Media*, 6.

12. "Whale."

13. H. E. Bowser, agent to P. T. Barnum, to George H. Newton, August 9, 1880, George H. Newton Papers, Collection 197, MC 87.4, MSM.

14. "Jonah's Friend."

15. "Only Manager."

16. George H. Newton to Warren Newton, April 5, 1881, George H. Newton Papers, Collection 197, MSM.

17. Unidentified newspaper clipping from the George H. Newton Papers, Collection 197, MSM.

18. Fred J. Engelhardt to George H. Newton, May 25, 1881, MC 87.11, Collection 197, MSM.

19. Engelhardt to Newton, May 25, 1881.

20. A published account of the whale's rehabilitation at the camp in Michigan was widely reprinted in newspapers in 1886 and affirms the account Engelhardt offered through his letters. "Traveled Whale."

21. "General Mention."

22. "Wonderful Lady Rider."

23. "Captain Paul Boyton, the great nautical adventurer, commanding the whaling crew, will be in the whale pavilion daily to receive and entertain visitors, and also display the wonderful life saving dress and other paraphernalia with which he makes his lonely voyages." "Monster Whale."

24. "Fred. Engelhardt's monster whale." "Traveled Whale."

25. On cars and trains as media, see Marshall McLuhan, *Understanding Media*. On moving walkways as media, see Huhtamo, "(Un)walking at the Fair." On brainwaves as media, see Kahn, *Earth Sound Earth Signal*. On insect swarms as media, see Parikka, *Insect Media*. Dana Luciano is also responsible for other creative and important readings of natural substances as media: Luciano, "Sacred Theories of Earth." The *Los Angeles Review of Books*' series on "Speaking Substances," drawn from an MLA panel of that title, also reads rocks, bodies, oil, and ice as ecomedia. Luciano, "Speaking Substances"; Blum, "Speaking Substances"; Schuller, "Speaking Substances"; Jones, "Oil."

26. "Oil itself is a medium that fundamentally supports all modern media forms concerned with what counts as culture—from film to recorded music, novels, magazines, photographs, sports, and the wikis, blogs, and videography of the internet." LeMenager, *Living Oil*, 6. LeMenager, "Infrastructure, Always and Again."

27. Shukin, *Animal Capital*, 20.

28. Shukin, *Animal Capital*, 108.

29. Jue, *Wild Blue Media*, 2; Alaimo, "Jellyfish Science, Jellyfish Aesthetics"; Alaimo, "New Materialisms, Old Humanisms, or, Following the Submersible"; Alaimo, "Violet-Black"; Blum, *View from the Masthead*; Blum, *News at the Ends of the Earth*.

30. Bolter and Grusin, *Remediation*.

31. Bolter and Grusin, *Remediation*, 45.

32. Bolter and Grusin, *Remediation*, 5, 21.

33. "Whale."

34. "Birds and Beasts." A report from the Chicago *Daily Inter Ocean* described the police as "Pinkertonian"; "Whale."

35. See especially chapter 2 of Samantha Frost's account of the porosity of cell boundaries as a defining trait of the "biocultural creature." Frost, *Biocultural Creatures*.

36. Harvey, "Portal of Touch."

37. Jenner, "Follow Your Nose?," 349. Important works in the history of smell include Corbin, *Foul and the Fragrant*; and Classen, Howes, and Synnott, *Aroma*. For a recent historiography of smell, see Reinarz, *Past Scents*.

38. Jay, "In the Realm of the Senses," 313.

39. "'Right' Here."

40. "Mammoth Proportions."

41. "Whale."

42. George H. Newton to Warren Newton, January 3, 1881, Collection 197, MSM.

43. George H. Newton to Warren Newton, January 4, 1881, Collection 197, MSM.

44. "Railrider Still Raiding Ohio."

45. Jenner, "Follow Your Nose?," 336.

46. The widely distributed documentary film *Blackfish* coalesced animal rights' protest around the exhibition of live orcas at sites such as SeaWorld.

47. Bergson, *Laughter*, 77.

48. "Casualties."

49. "Whale."

50. See especially John Patrick Leary's entry on "innovation," which traces the term's proliferation in twenty-first-century neoliberal capitalism to nineteenth-century concepts of "progress." Leary, *Keywords for the Age of Austerity*, 114–19.

51. Norris, *Octopus*.

52. Star, "Ethnography of Infrastructure," 382.

53. Jessica Hurley and Jeffrey Insko effectively discuss the invisibility/visibility paradigm in their introduction to the special issue of *American Literature* on "The Infrastructure of Emergency." Hurley and Insko, "Introduction."

54. "Railroads."

55. Susan Nance, *Entertaining Elephants*, 138–74.

56. Historian Richard White writes about how this progress-based myth has attended contemporary and historical accounts of the development of the transcontinental railroad, and he attempts to revise that history by exposing the chaos and disorganization of the railroads' development in *Railroaded*.

57. "Distinguished Visitor."

58. Margaret Cohen writes that one of the effects of technical detail in maritime fiction is to establish the sailor's "practical resourcefulness," or "practical reason, . . . the philosophical term for the intelligence distinguishing people who excel in the arts of action." Cohen, *Novel and the Sea*, 2.

59. The public historian Daniel Gifford—great-great-grandson of the captain who brought the *Progress* to Chicago—offers a comprehensive and lively narrative history of the *Progress*'s life and strange afterlife. Gifford, *Last Voyage*.

60. "Whales at the Fair."

61. "Whaling Big Boat."

62. "Death Calls Ossian Guthrie"; Gifford, *Last Voyage*.

63. "Nautical Notes."

64. Gifford, *Last Voyage*.

65. "Ancient Mariner."
66. "Lake Shipping News."
67. "Arctic Sea Trophies."
68. "Bark Progress Here."
69. "Arctic Sea Trophies."
70. Stevens, *Six Months at the World's Fair*, 20.
71. "Fate of the Whaler Progress."
72. "Whaler Ditched."
73. Gifford reveals that there were multiple failed attempts back in New Bedford to bring the vessel "home"; though forgotten in Chicago, it was somewhat worried over in New Bedford. Gifford, *Last Voyage*, 203–24.
74. "Whaling Exhibit Is Purchased"; Legge, "Last Voyage of the Progress."
75. "Whaler Ditched."
76. "Whaler Coming Back from Chicago."
77. Gifford, *Last Voyage*, 164.
78. Daniel Gifford also offers a sustained interpretation of the *Progress* as a "larger metaphor for the continued decline and decay of whaling." Gifford, *Last Voyage*, 215.
79. Gifford, *Last Voyage*, 83–84.
80. *Industrial Chicago*.
81. "Death Calls Ossian Guthrie."
82. "Whaler Progress Sunk."
83. "Exit the Progress."
84. "Ancient Mariner"; Gifford, *Last Voyage*, 159–60.
85. Gifford, *Last Voyage*, 115–53.
86. "Waterways and Water Pageants at the Fair."
87. "Whaler Ditched."
88. Rydell, *All the World's a Fair*, 40.
89. Rydell, *All the World's a Fair*, 63–64.
90. On Indigenous, Soviet, and US commercial whaling in the Arctic, see Demuth, *Floating Coast*.
91. Stevens, *Six Months at the World's Fair*, 20.
92. Gifford, *Last Voyage*, 131.
93. McLuhan, *Understanding Media*, 224.
94. Dona Brown offers an account of the transformation of the Nantucket whaling economy into a tourist economy in chapter 5 of her excellent book on New England tourism, *Inventing New England*, "'Manufactured for the Trade,'" 105–34.

Chapter 4

1. Ashley, *Yankee Whaler*, 117–18.
2. Conn, *Museums and American Intellectual Life*, 9.
3. Rosaldo, "Imperialist Nostalgia," 107–8.
4. For a more thorough and focused treatment of masculinity in the nineteenth-century whaling industry, see Schell, *Bold and Hardy Race of Men*.
5. Daggett, "Petro-Masculinity," 28.
6. Whyte, "Dakota Access Pipeline"; Whyte, "Indigenous Climate Change Studies"; Estes, *Our History Is the Future*.

7. O'Brien, *Firsting and Lasting*, 143.

8. The narrative of the dying Yankee whaling industry in some ways resembles the narrative arc of the myth of the "vanishing Indian," which Philip Deloria defines as an ideology that "proclaimed it foreordained that less advanced societies should disappear in the presence of those more advanced." Deloria, *Playing Indian*, 64. The loss of the Yankee whaling industry appears in the early twentieth-century accounts I describe here as a cultural formation becoming "extinct": one whose extinction is mourned but ultimately accepted because it gave way to the new and modern, fossil-fueled power. But there is an essential difference between the two "vanishing" narratives. The vanishing Indian narrative whitewashes a genocide. The bygone Yankee whaling narrative whitewashes an energy transition—bound up, to be sure in settler colonialism and mass violence, but not directly equivalent to genocide. It is for this reason that I chose in chapter 4 to consider the "bygone Yankee whaling" narrative through the lens of Jean O'Brien's concept of "firsting and lasting," which touches on the myth of the vanishing Indian but addresses other practices of settler historiography and storytelling in terms more readily available to the context of Yankee whaling.

9. O'Brien, *Firsting and Lasting*, 134–39.

10. Barnard, "Cod and the Whale."

11. Webster and Erickson, "Last Word?"

12. Tønnessen and Johnsen, *History of Modern Whaling*.

13. Warrin, *So Ends This Day*; Shoemaker, *Living with Whales*; Shoemaker, *Native American Whalemen*.

14. Warrin, *So Ends This Day*, 265.

15. Hohman, *American Whaleman*, 301.

16. Lowenthal, *Heritage Crusade*, 128.

17. Heritage is often associated with populist ideologies, but as Lowenthal points out, "heritage normally goes with privilege: elites usually own it, control access to it, and ordain its public image." The case of early twentieth-century New Bedford whaling heritage is no different. Lowenthal, *Heritage Crusade*, 90.

18. *Presentation of the Whaleman Statue*, 4; Lindgren, "'Let Us Idealize Old Types of Manhood.'"

19. *Presentation of the Whaleman Statue*, 10.

20. *Presentation of the Whaleman Statue*, 11.

21. On Indigenous people in the New England whaling industry, see Shoemaker, *Living with Whales*. See also the ongoing work of Jason Mancini in the Indian Mariners Project at https://indianmarinersproject.com.

22. *Presentation of the Whaleman Statue*, 16–17.

23. *Presentation of the Whaleman Statue*, 39–40.

24. Spiro, *Defending the Master Race*, 4–5; Taylor, *American Conservation Movement*, 168.

25. Taylor, *American Conservation Movement*, 221.

26. Taylor, *American Conservation Movement*, 52.

27. Gilmore, *Manhood in the Making*; Slotkin, *Regeneration through Violence*.

28. Taylor, *American Conservation Movement*, 67.

29. Spiro, *Defending the Master Race*, 4.

30. Spiro, *Defending the Master Race*, 73.

31. Even Aldo Leopold, widely revered as an early advocate for ecological science and environmentalism, formalized practices of wildlife management and the theory of the "conservation ethic" with elitist sport hunters in mind; in his view, hunters through thoughtful and informed culls could promote ecological balance in a given landscape. The type of hunting he envisioned was the elitist sport hunting promoted by the Boone and Crockett Club, of which he was a longtime member. Spiro, *Defending the Master Race*, 82.

32. Spiro, *Defending the Master Race*, 100–116.

33. Spiro, *Defending the Master Race*.

34. Spiro, *Defending the Master Race*; Taylor, *American Conservation Movement*; Powell, *Vanishing America*; Tompkins, "Sierra Club Says It Must Confront the Racism of John Muir."

35. Tønnessen and Johnsen, *History of Modern Whaling*, 3; Roman, *Whale*.

36. Demuth, *Floating Coast*.

37. Ashley, *Yankee Whaler*, 30.

38. William McFee wrote maritime narratives and is notable, according to Bert Bender, because he closely associated himself with the new era of steam, rather than writing nostalgically about the obsolescing sailing era. Bender, *Sea-Brothers*, 136–39.

39. McFee, "Foreword," ix.

40. McFee, "Foreword," x.

41. McFee, "Foreword," x.

42. Melville and Parker, *Moby-Dick*, 164.

43. Nicholas, *VC*.

44. Barrett, "Picturing a Crude Past."

45. Barnard, "Cod and the Whale."

46. McFee, "Foreword," x.

47. "Articles of Incorporation from the Commonwealth of Massachusetts for The Whaling Film Corporation," MSS39: Whaling Film Corporation Records, NBWM.

48. On Ashley's painting and other art and craftwork in the context of early twentieth-century New Bedford, see Nicole Williams, "'Whalecraft.'"

49. Ashley, *Yankee Whaler*, 118.

50. Hodkinson Pictures, *Elmer Clifton's Down to the Sea in Ships*, 12. MSS39: Whaling Film Corporation Records, NBWM.

51. Hodkinson Pictures, "Actor Starts Whale Hunt Unprepared."

52. McKee, *Scrimshaw Sperm Whale Tooth*, Collections Research Center, Mystic Seaport Museum.

53. "'Interviewed by Questionnaire' [Typescript of an Oral History of the Making of Down to the Sea in Ships, c. 1922]," 1. MSS39: Whaling Film Corporation Records, NBWM.

54. "'Interviewed by Questionnaire,'" 3.

55. LeMenager, *Living Oil*, 6.

56. Blair, "Logbook of the Gaspe," 10. ODHS 0603: Logbook Collection, NBWM.

57. Martingale [Sleeper], *Salt Water Bubbles*.

58. Ashley, "Clifford Ashley to Elmer Clifton," ca. 1922.

59. The *Progress*, the whaling ship that had been exhibited at the Columbian Exposition in Chicago, whose story I trace in chapter 3, could be counted the third whaling ship that catalyzed New Bedford whaling preservation efforts. This is the argument

made by Daniel Gifford, who observes that "less than ten months after New Bedford read about the final conflagration of the *Progress*, Ellis Howland stood at a podium and demanded that New Bedford create a museum to whaling." Gifford, *Last Voyage*, 217.

60. Coined in 1996 in a letter to the journal *Nature*, the term "endling" has gained some traction in environmental activism. Webster and Erickson, "Last Word?"; Jørgensen, "Naming and Claiming the Last."

61. Woodward, "Cruise of Wanderer May Be Last from This Port."

62. "Wanderer's Crew All Safe on Shore."

63. A short description of Green's involvement with the *Morgan* appears in a book that offers the ship's biography: Leavitt, *Charles W. Morgan*, 78–81. "Col. Green to Spend $100,000 Restoring Last of Whaling Barks."

64. "Col. Green to Spend $100,000 Restoring Last of Whaling Barks."

65. "Old Whaler Dedicated as Memorial to Industry."

66. Coryell, "Green, Hetty."

67. Bedell, *Colonel Edward Howland Robinson Green*.

68. Bedell, *Colonel Edward Howland Robinson Green*, 38–39.

69. "Tax Hearing Turns on Green's Hobbies."

70. "Old Whaler Dedicated as Memorial to Industry."

71. In his meticulously researched article on the New Bedford Whaling Museum, James Lindgren discovered that the *Morgan* brought in many, many more visitors than did the whaling museum at the Old Dartmouth Historical Society during the same years: in the year after June 1, 1927, according to Lindgren, "189,851 people came aboard, over twenty-five times the gate at the ODHS museum." Lindgren, "'Let Us Idealize Old Types of Manhood.'"

72. "Whaler Preserved in Concrete Recalls New Bedford Romance."

73. "Whaler Preserved in Concrete Recalls New Bedford Romance."

74. Green had hoped to expand the site of the *Morgan* to include a miniature model whaling village, but the exhibition of the *Morgan* ended with Green's death in 1936. Green funded Whaling Enshrined while he was alive, and the venture thrived. But Green failed to establish a trust for the project after his death, and the *Morgan* entered yet another crisis in the mid-1930s. "Memorial to Whaling."

75. Ashley, *Yankee Whaler*, 56.

76. Conn, *Museums and American Intellectual Life*, 152.

77. Simpson and Szeman, "Impasse Time," 81.

78. Daggett, "Petro-Masculinity," 32.

79. Daggett, "Petro-Masculinity"; Boyer and Howe, "Timothy Mitchell."

80. For Arby's, see Ingraham, "Entire Coal Industry"; for Walt Disney World, see Wood and Tong, "Counting Up American Coal Jobs."

81. David, "Town's Nostalgia for Seafaring Past."

82. Plumer and Tankersley, "Trump Picks Economic Winners."

83. Lipton, "Coal Industry Is Back."

84. Quoted in Lipton, "Coal Industry Is Back."

85. Skinner, *Behavior of Organisms*; Lerman and Iwata, "Prevalence of the Extinction Burst and Its Attenuation during Treatment"; Harris, Pentel, and LeSage,

"Prevalence, Magnitude, and Correlates of an Extinction Burst in Drug-Seeking Behavior in Rats."

86. Rosaldo, "Imperialist Nostalgia," 110.

Chapter 5

1. Some of the biographies and critical texts that contributed to the first wave of the Melville Revival are Weaver, *Herman Melville*; Mumford, *Herman Melville*; and Matthiessen, *American Renaissance*. Some of the most insightful accounts of the Melville Revival include: Lauter, "Melville Climbs the Canon"; Spark, *Hunting Captain Ahab*; Insko, "'All of Us Are Ahabs'"; Aronoff, "Cultures, Canons, and Cetology"; and Buell, *Dream of the Great American Novel*.

2. Renker, *Origins of American Literature Studies*; Vanderbilt, *American Literature and the Academy*; Shumway, *Creating American Civilization*; Buell, *Dream of the Great American Novel*.

3. Lauter, "Melville Climbs the Canon"; Spanos, *Errant Art of Moby-Dick*; Spark, *Hunting Captain Ahab*.

4. Aronoff, "Cultures, Canons, and Cetology," 186.

5. In my reading of Rockwell Kent's *Moby-Dick* and its place in US literary and intellectual history, I build on the work of, among others, Elizabeth Schultz and Angela Miller, who have also presented Kent's illustration as powerful works of Melville interpretation. Schultz, *Unpainted to the Last*; Miller, "Reading Ahab."

6. "Just as energy became tightly bound by the governing logic of work, so too work increasingly came to be governed through the metaphors and physics of energy. The energy-work bindings were laced tight in the nineteenth century, with the purported discovery of energy and its service to Western fossil-fueled imperialism. The Western epistemology of energy attached fuel systems to the gospel of labor and its veneration of productivity. The energy-work nexus was so friendly to the spread of fossil capital, so conducive to concealing its violence, and so minutely sutured as to leave little trace of its contingent pairing. The intertwining of energy and the Western ethos of dynamic, productive work was produced as cosmic truth." Daggett, *Birth of Energy*, 5.

7. It is important to note that industrial whaling at large did *not* end at this moment; it has not ended. The industrial slaughter of whales actually accelerated in the twentieth century and was carried out mainly by British and Norwegian whaling fleets, a fact that the commemorators I chronicle in chapter 4 note in order to make the case for the particular (white) Americanness of the nineteenth-century whaling industry. Tønnessen and Johnsen, *History of Modern Whaling*.

8. Schultz also found that Kent's *Moby-Dick* illustrations registered "the schism and tension in Kent's philosophical convictions and his actions— . . . his involvement in capitalism versus his commitment to socialism. Schultz, *Unpainted to the Last*, 29.

9. Weaver, *Herman Melville*, 134; Mumford, *Herman Melville*, 42.

10. Weaver, *Herman Melville*, 154.

11. Weaver, *Herman Melville*, 33; Lauter, "Melville Climbs the Canon," 8.

12. Daggett, *Birth of Energy*.

13. Weaver, *Herman Melville*, 28.

14. Mumford, *Herman Melville*, 196.

15. Linthicum, Relford, and Johnson, "Defining Energy in Nineteenth-Century Native American Literature"; Szeman and Boyer, "Introduction."

16. Mumford, *Technics and Civilization*, 156; Malm, *Fossil Capital*.

17. Mumford, *Technics and Civilization*, 158.

18. Mumford, *Herman Melville*, 64–65.

19. Mumford, *Herman Melville*, 294.

20. Eric Aronoff argues in his insightful account of Mumford's Melville biography that the good feature of the world that produced Melville's talent through *Moby-Dick* is owed to "regionalism" and to Melville's access to a sense of regional New England identity. I argue for a different locus of meaning in Mumford's biography, centered more on technological and energic terms. Aronoff, "Cultures, Canons, and Cetology."

21. Mumford, *Herman Melville*, 188.

22. Mumford, *Herman Melville*, 306.

23. Mumford, *Herman Melville*, 48.

24. Lawrence, *Studies in Classic American Literature*, 160.

25. Jeffrey Insko also noted that "Olson . . . places the novel in relation to the growth of the petroleum industry." Insko, "'All of Us Are Ahabs,'" 31.

26. Olson et al., *Collected Prose*, 17–18.

27. William A. Kittredge to Rockwell Kent, September 16, 1926, Rockwell Kent Papers, Series 1: Alphabetical Files, Archives of American Art; Badaracco, *American Culture and the Marketplace*, 54.

28. Frost, "Masque of Mercy," 495.

29. Schultz, *Unpainted to the Last*, 43.

30. Of these scholarly accounts, Schultz's is the most comprehensive, arguing that Kent's work "embodies that fundamental dichotomy in Melville's novel between the physical, the concrete, and the finite on the one hand, and the metaphysical, the abstract, and the infinite on the other." Schultz, *Unpainted to the Last*, 52. Angela Miller's excellent account notes that Kent's black-and-white illustrations resonate with Ahab's worldview, which she characterizes as self-mythologizing and even totalitarian, in light of Cold War anxieties that colored the novel's reception later in the twentieth century. Miller, "Reading Ahab," 223–45. See also Abrams, "Illuminated Critique."

31. Benson, *Printed Picture*, 6, 8–10.

32. Mumford, *Technics and Civilization*, 119, 158.

33. Yaeger, "Editor's Column," 305.

34. West, "Rockwell Kent," 114.

35. Rockwell Kent, *It's Me O Lord*, 308.

36. Leighton, *Wood Engraving*, 17.

37. Schultz, *Unpainted to the Last*, 30.

38. Bluemel, "Rural Modernity," 245.

39. Kent, "Workers of the World, Unite!"

40. Ruskin, *Ariadne Florentina*, 29–30, 67.

41. Jones, "Print Nostalgia."

42. Not all critics misidentify Kent's illustrations for *Moby-Dick*. I am particularly grateful to Cecilia Esposito and Jake Milgram Wien, who, through published scholarship and private correspondence, have shaped my understanding of Kent's drawing

and print practice and the long-standing confusion around the medium of his work. Cecilia Esposito, Director/Curator, Plattsburgh State Art Museum, in discussion with the author, January 2017. For corrections to prior misidentifications of Kent's media, see Wien, "Rockwell Kent." Elizabeth Schultz reflects on the medium of Kent's illustrations as drawing at length in Schultz, "The Illustrations of Rockwell Kent and His Followers," in *Unpainted to the Last*, 27–63.

43. As reworking of other images, Kent's skeuomorphic line drawings could also be understood as remediations, following Bolter and Grusin. But the term "skeuomorph" affords me a more precise term for describing the ideological work of Kent's illustrations and, in particular, their relationship to the past. Bolter and Grusin, *Remediation*.

44. Boym, *Future of Nostalgia*, xiii.

45. Leighton, *Wood Engraving*, 9.

46. Benson, *Printed Picture*, 6–26, 210–71; Brown, *Beyond the Lines*, 236.

47. Rockwell Kent, *This is My Own*, 54, 57, emphasis added.

48. "Skeuomorph, n.," *OED Online*, Oxford University Press, 2022, www.oed.com/view/Entry/180780.

49. Dan O'Hara, "Skeuomorphology and Quotation," 283. Most accounts of skeuomorphism come from one of two fields: first, archaeology, where skeuomorphs offer clues into technological history and the relative values of ancient crafts objects; and second, digital design, where skeuomorphism is deployed as a strategy to help users navigate a digital environment by embedding signs of the material world in it. A Bronze Age ceramic vessel wrought in the shape and texture of a woven basket is a skeuomorph. The "folders" in which you save your digital "files" on your computer are skeuomorphs. So, too, are Kent's woodcut-style ink drawings. Carl Knappett, "Photographs, Skeuomorphs and Marionettes," 97–117.

50. Henning, "New Lamps for Old."

51. Schrey, "Analogue Nostalgia and the Aesthetics of Digital Remediation," 36.

52. Huhtamo and Parikka, *Media Archaeology*; Parikka, *What Is Media Archaeology?*

53. Huhtamo, "Dismantling the Fairy Engine," 28.

54. Carlson, "Untold Story."

55. Chatfield, "Apple"; Brownlee, "Most Hated Design Trend."

56. Traxel, *American Saga*, 197, 200–210; Chunikhin, "At Home among Strangers."

57. Kent, "How I Make a Wood Cut."

58. Kent used the phrase "multiple originals" in another work to signal his belief that print was the most accessible form of art: "It is now realized, as perhaps never before, that not only does art properly concern itself with the expression of universal values but that its appeal must be directed to humanity at large. In the new spirit that has come to art, no form of art can be more effective than printmaking. It is the art of multiple originals." Quoted in Jones, *Prints of Rockwell Kent*, xii.

59. Kent, *Fifty Prints Exhibited by the Institute*, xii.

60. Sloan, *New York Scene*, 431.

61. Kent, *This Is My Own*, 13.

62. Kent, *This is My Own*, 13; Kent, *It's Me O Lord*, 275, 273.

63. Corey Ford's most successful parody was [June Triplett, pseud.], *Salt Water Taffy; or, Twenty Thousand Leagues away from the Sea*, a parody of the sailing adventure narrative by Joan Lowell, *Child of the Deep*. On Corey Ford, see Day, "Corey Ford,"

147–51. For a more recent critical account that charges Kent with political inconsistency, see Miller, "Reading Ahab," 223–45.

64. John Riddell [Corey Ford], "N by G," in *In the Worst Possible Taste*; and Kent, *N by E*.

65. Riddell, "N by G," 54.

66. Traxel, *American Saga*, 116–20.

67. Riddell, "N by G," 55, 56, 67.

68. Riddell, "N by G," 58.

69. Leighton, *Wood Engraving*, 176. Leighton refers to the beginning of the Great Depression as the "slump." On page 10 of the book, she calls it the "slump of 1931."

70. Cassidy, *Marsden Hartley*, 201–2. Depicting Maine laborers was a practice of modernist primitivism, according to Cassidy, whose work fills a significant scholarly lacuna by reading race back into depictions of the North Atlantic folk in *Marsden Hartley*, 169–212.

71. Kent, *It's Me O Lord*, 117–58 (quotation at 122). On this period, see also Traxel, *American Saga*.

72. Kent, *It's Me O Lord*, 120.

73. Jonathan Weinberg offers an extended analysis of the images from the American Car and Foundry series, although he focuses on the hypermasculinity and homoeroticism of the images. See Weinberg, "I Want Muscle," 115–34. Wien also offers a concise history of the American Car and Foundry commission in Wien, *Mythic and the Modern*, 118–20.

74. Weinberg, *Male Desire*, 71.

75. Lears, *No Place of Grace*, 95.

76. Jones, *Prints of Rockwell Kent*, xiv. Ione Robinson describes Kent's workshop in her memoir, *A Wall to Paint On*, 30–34.

77. Rockwell Kent to Marques E. Reitzel, n.d., quoted in Jones, *Prints of Rockwell Kent*, xiii.

78. Cassidy, *Marsden Hartley*, 66–67.

79. Blum, *View from the Masthead*; Blum, "Prospect of Oceanic Studies"; Cohen, *Novel and the Sea*; Parsons, "'Careful Disorderliness.'"

Epilogue

1. Smith, "Thou Uncracked Keel"; Leavitt, *Charles W. Morgan*.

2. Leavitt, *Charles W. Morgan*, 87.

3. The line comes from Robert Lowell's poem about the Fifty-Fourth Regiment Memorial monument on the Boston Common: a bronze relief sculpture that commemorates the all-Black Fifty-Fourth Regiment of the Union Army and its white leader, Colonel Robert Gould Shaw. Lowell wrote about the monument in the poem "For the Union Dead" during the Civil Rights struggles of the 1960s. Lowell, *Collected Poems*.

4. Jackson, "Rethinking Repair," 221.

5. Jackson, "Rethinking Repair," 226.

6. White interview.

7. Melville and Parker, *Moby-Dick*, 39.

8. Melville and McCall, "Benito Cereno," 65.

9. Warren, "Remarks."

10. Gelles, "The Texas Group Waging a National Crusade against Climate Action."
11. Mitchell, "Remarks."
12. I discuss "extractivist nostalgia" in chapter 4.
13. Davenport and Friedman, "Biden Administration Approves Nation's First Major Offshore Wind Farm."
14. Lennon, "America's Leader in Offshore Wind."
15. McCarron, "Offshore Wind Company Lays Final Cable at Barnstable Beach. What's Next?"
16. Mulligan, "New Bedford Offshore Wind Business Blows Its Stack."
17. Territorial Agency, Palmesino, and Rönnskog, "Radical Conservation," 338.
18. Territorial Agency, Palmesino, and Rönnskog, "Radical Conservation," 350.
19. Insko, "On Dismantling," 142.
20. Insko, "On Dismantling," 146.
21. Insko, "On Dismantling," 146.
22. Odell, *How to Do Nothing*, 190.
23. Lawrence interview.

BIBLIOGRAPHY

Manuscript and Archival Collections

Mystic, CT
    Mystic Seaport Museum, Collections Research Center
        Manuscript Collection 19: Records of the Bark Charles W. Morgan
        Manuscript Collection 87: Capt. John W. Carter Papers
        Manuscript Collection 197: George H. Newton Papers
Nantucket, MA
    Nantucket Historical Association
        Collection MS 155: Steamboats
        Collection MS 236: Nantucket Lodgings
        Collection MS 397: Tourist Guides
            Vertical Files
New Bedford, MA
    New Bedford Whaling Museum Research Library
        MSS39: Whaling Film Corporation Records
        ODHS 0603: Logbook Collection
        MSS61: Akin Family Papers
Washington, DC
    Archives of American Art
        Rockwell Kent Papers
            Vertical Files

Primary and Secondary Sources

"Aboriginal Subsistence Whaling in the Arctic." International Whaling Commission. Accessed November 29, 2018. https://iwc.int/aboriginal.

Abrams, Matthew Jeffrey. "Illuminated Critique: The Kent *Moby-Dick*." *Word and Image* 33, no. 4 (2017): 376–91.

Acland, Charles R. "Introduction: Residual Media." In *Residual Media*, xiii–xxvii. Minneapolis: University of Minnesota Press, 2007.

———, ed. *Residual Media*. Minneapolis: University of Minnesota Press, 2007.

Alaimo, Stacy. "Jellyfish Science, Jellyfish Aesthetics: Posthuman Reconfigurations of the Sensible." In *Thinking with Water*, edited by Cecilia Chen, Janine MacLeod, and Astrida Neimanis, 139–64. Montreal: McGill-Queen's University Press, 2013.

———. "New Materialisms, Old Humanisms, or, Following the Submersible." *NORA: Nordic Journal of Feminist and Gender Research* 19, no. 4 (2011): 280–84.

———. "Violet-Black." In *Prismatic Ecology: Ecotheory beyond Green*, ed. Jeffrey Jerome Cohen, 233–51. Minneapolis: University of Minnesota Press, 2013.

Allen, Thomas M. *A Republic in Time: Temporality and Social Imagination in Nineteenth-Century America*. Chapel Hill: University of North Carolina Press, 2008.

Anand, Nikhil, Akhil Gupta, and Hannah Appel, eds. *The Promise of Infrastructure*. Durham, NC: Duke University Press, 2018.

"An Ancient Mariner." *Daily Inter Ocean* (Chicago), July 19, 1892.

"Arctic Sea Trophies." *Daily Inter Ocean* (Chicago), July 28, 1892.

Aron, Cindy Sondik. *Working at Play: A History of Vacations in the United States*. Oxford: Oxford University Press, 2001.

Aronoff, Eric. "Cultures, Canons, and Cetology: Modernist Anthropology and the Form of Culture in Lewis Mumford's Herman Melville." *ESQ: A Journal of the American Renaissance* 58, no. 2 (2012): 185–217.

Ashley, Clifford. *The Yankee Whaler*. New York: Dover, 1991.

Austin, Dan, and Sean Doerr. *Lost Detroit: Stories behind the Motor City's Majestic Ruins*. Charleston, SC: History Press, 2010.

Badaracco, Claire. *American Culture and the Marketplace: R. R. Donnelley's Four American Books Campaign, 1926–1930*. Washington, DC: Library of Congress, 1992.

Baker, Jennifer J. "Hawthorne's Picturesque at Home and Abroad." *Studies in Romanticism* 55, no. 3 (2016): 417–44.

"Bark Progress Here: The Old Whaler Lies in Anchor in the Basin." *Chicago Daily Tribune*, July 28, 1892.

Barnard, John Levi. "The Cod and the Whale: Melville in the Time of Extinction." *American Literature* 89, no. 4 (2017): 851–79.

Barnard, John Levi, and Stephanie Foote. "The Ruins of the Future." *Resilience: A Journal of the Environmental Humanities* 8, no. 3 (2021): 1–14.

Barney, Darin. "Pipelines." In *Fueling Culture: 101 Words for Energy and Environment*, edited by Imre Szeman, Jennifer Wenzel, and Patricia Yaeger, 267–70. New York: Fordham University Press, 2017.

Barnett, Lydia. *After the Flood: Imagining the Global Environment in Early Modern Europe*. Baltimore: Johns Hopkins University Press, 2019.

Barrett, Ross. "Picturing a Crude Past: Primitivism, Public Art, and Corporate Oil Promotion in the United States." In *Oil Culture*, edited by Ross Barrett and Daniel Worden, 43–68. Minneapolis: University of Minnesota Press, 2014.

Barrett, Ross, and Daniel Worden, eds. *Oil Culture*. Minneapolis: University of Minnesota Press, 2014.

Beard, George M. *American Nervousness: Its Causes and Consequences*. New York: G. P. Putnam's Sons, 1881.

Bedell, Barbara Fortin. *Colonel Edward Howland Robinson Green and the World He Created at Round Hill*. South Dartmouth, MA: The Author, 2003.

Bender, Bert. *Sea-Brothers: The Tradition of American Sea Fiction from* Moby-Dick *to the Present*. Philadelphia: University of Pennsylvania Press, 1990.

Benson, Richard. *The Printed Picture*. New York: Museum of Modern Art, 2008.

Berardi, Franco. "Exhaustion." In *Fueling Culture: 101 Words for Energy and Environment*, edited by Imre Szeman, Jennifer Wenzel, and Patricia Yaeger, 155–57. New York: Fordham University Press, 2017.

Berger, Jason. *Antebellum at Sea: Maritime Fantasies in Nineteenth-Century America*. Minneapolis: University of Minnesota Press, 2012.

Bergson, Henri. *Laughter: An Essay on the Meaning of the Comic*. London: Macmillan, 1913.

Bernstein, Robin. *Racial Innocence: Performing American Childhood from Slavery to Civil Rights*. New York: New York University Press, 2011.

Black, Brian. *Petrolia: The Landscape of America's First Oil Boom*. Baltimore: Johns Hopkins University Press, 2000.

Bleumel, Kristin. "Rural Modernity and the Wood Engraving Revival in Interwar England." *Modernist Cultures* 9, no. 2 (2014): 233–59.

Blum, Hester. *The News at the Ends of the Earth: The Print Culture of Polar Exploration*. Durham, NC: Duke University Press, 2019.

———. "The Prospect of Oceanic Studies." *PMLA* 125, no. 3 (2010): 670–77.

———. "Speaking Substances: Ice." *Los Angeles Review of Books*, March 21, 2016.

———. *The View from the Masthead: Maritime Imagination and Antebellum American Sea Narratives*. Chapel Hill: University of North Carolina Press, 2008.

Bockstoce, John R. *Whales, Ice, and Men: The History of Whaling in the Western Arctic*. Seattle: University of Washington Press in association with the New Bedford Whaling Museum, Massachusetts, 1986.

Bolter, J. David, and Richard A. Grusin. *Remediation: Understanding New Media*. Cambridge, MA: MIT Press, 1999.

Bolton, Francis. *London Water Supply, Including a History and Description of the London Waterworks, Statistical Tables, and Maps*. London: William Clowes and Sons, 1888.

Bonneuil, Christophe, and Jean-Baptiste Fressoz. *The Shock of the Anthropocene*. Translated by David Fernbach. London: Verso, 2017.

Borowy, Iris, and Matthias Schmelzer, eds. *History of the Future of Economic Growth: Historical Roots of Current Debates on Sustainable Degrowth*. New York: Routledge, 2018.

Boyer, Dominic, and Cymene Howe. "Ep. #57—Timothy Mitchell." February 16, 2017. In *Cultures of Energy*, produced by Mingomena Media, podcast, 76 min. https://web.archive.org/web/20180626194814/http://culturesofenergy.com/ep-57-timothy-mitchell/.

Boym, Svetlana. *The Future of Nostalgia*. New York: Basic Books, 2001.

Bramen, Carrie Tirado. "The Urban Picturesque and the Spectacle of Americanization." *American Quarterly* 52, no. 3 (2000): 444–77.

Brown, Dona. *Inventing New England: Regional Tourism in the Nineteenth Century*. Washington, DC: Smithsonian Institution Press, 1995.

Brown, Joshua. *Beyond the Lines: Pictorial Reporting, Everyday Life, and the Crisis of Gilded Age America*. Berkeley: University of California Press, 2002.

Brownlee, John. "The Most Hated Design Trend Is Back." *Fast Co. Design*, October 3, 2014. www.fastcodesign.com/3036347/the-most-hated-design-trend-is-back.

Buell, Frederick. "Energy Systems." In *Fueling Culture: 101 Words for Energy and Environment*, edited by Imre Szeman, Jennifer Wenzel, and Patricia Yaeger, 140–44. New York: Fordham University Press, 2017.

Buell, Lawrence. *The Dream of the Great American Novel*. Cambridge, MA: Belknap Press of Harvard University Press, 2014.

———. *Writing for an Endangered World: Literature, Culture, and Environment in the U.S. and Beyond*. Cambridge, MA: Belknap Press of Harvard University Press, 2001.

Burleigh [Matthew Smith Hale]. *The Old Colony Railroad: Its Connections, Popular Resorts, and Fashionable Watering-Places*. Boston: Rand, Avery, 1877.

Burnett, D. Graham. *The Sounding of the Whale: Science and Cetaceans in the Twentieth Century*. Chicago: University of Chicago Press, 2012.

Carlson, Nicholas. "The Untold Story of How Steve Jobs Reintroduced His Signature Design Style to Apple." *Business Insider*, October 19, 2014. www.businessinsider.com/steve-jobss-signature-design-style-2014-10.

Casarino, Cesare. *Modernity at Sea: Melville, Marx, Conrad in Crisis*. Minneapolis: University of Minnesota Press, 2002.

Cassidy, Donna M. *Marsden Hartley: Race, Region, and Nation*. Hanover, NH: University Press of New England, 2005.

"Casualties: A Crash at Cincinnati." *St. Louis Globe-Democrat*, February 19, 1881.

Chatfield, Tom. "Apple: An End to Skeuomorphic Design?" *BBC Future*, May 9, 2013. www.bbc.com/future/story/20130509-apple-designs-break-from-the-past.

Chunikhin, Kirill. "At Home among Strangers: U.S. Artists, the Soviet Union, and the Myth of Rockwell Kent during the Cold War." *Journal of Cold War Studies* 21, no. 4 (2019): 175–207.

Classen, Constance, David Howes, and Anthony Synnott, eds. *Aroma: The Cultural History of Smell*. London: Routledge, 1994.

"Coal Oil." *White Cloud Kansas Chief*, April 4, 1861.

Cohen, Daniel, and Jane Marie Todd. *The Infinite Desire for Growth*. Princeton, NJ: Princeton University Press, 2018.

Cohen, Margaret. *The Novel and the Sea*. Princeton, NJ: Princeton University Press, 2010.

"Col. Green to Spend $100,000 Restoring Last of Whaling Barks." *Boston Daily Globe*, October 19, 1924.

Conn, Steven. *Museums and American Intellectual Life, 1876–1926*. Chicago: University of Chicago Press, 1998.

Conron, John. *American Picturesque*. University Park: Pennsylvania State University Press, 2000.

Cook, R. H. *Historical Notes of the Island of Nantucket and Tourist's Guide*. Nantucket, MA, 1871.

Corbin, Alain. *The Foul and the Fragrant: Odor and the French Social Imagination*. Cambridge, MA: Harvard University Press, 1986.

———. *The Lure of the Sea: The Discovery of the Seaside in the Western World, 1750–1840*. Translated by Jocelyn Phelps. Berkeley: University of California Press, 1994.

Coronil, Fernando. *The Magical State: Nature, Money, and Modernity in Venezuela*. Chicago: University of Chicago Press, 1997.

Coryell, Janet L. "Green, Hetty." *American National Biography Online*, February 2000. www.anb.org.ezp-prod1.hul.harvard.edu/articles/10/10-00679.html.

Cowperthwaite, Gabriela, dir. *Blackfish*. DVD. Los Angeles: Magnolia Home Entertainment, 2013.

Cronon, William. *Nature's Metropolis: Chicago and the Great West*. New York: W. W. Norton, 1992.

Daggett, Cara New. *The Birth of Energy: Fossil Fuels, Thermodynamics, and the Politics of Work*. Durham, NC: Duke University Press, 2019.

———. "Petro-Masculinity: Fossil Fuels and Authoritarian Desire." *Millennium: Journal of International Studies* 47, no. 1 (2018): 25–44.

Davenport, Coral, and Lisa Friedman. "Biden Administration Approves Nation's First Major Offshore Wind Farm." *New York Times*, May 11, 2021.

Davis, John. "Going Analog: Vinylphiles and the Consumption of the 'Obsolete' Vinyl Record." In *Residual Media*, edited by Charles R. Acland, 16–31. Minneapolis: University of Minnesota Press, 2007.

Davis, Lance E., Robert E. Gallman, and Karin Gleiter. *In Pursuit of Leviathan: Technology, Institutions, Productivity, and Profits in American Whaling, 1816–1906*. Chicago: University of Chicago Press, 1997.

Davis, Thomas S. "'Far from the Gulf Coast, but Near It, Too': Art, Attachment, and Deepwater Horizon." *Resilience: A Journal of the Environmental Humanities* 8, no. 3 (2021): 71–97.

Day, Patrick. "Corey Ford." In *American Humorists, 1800–1950*, edited by Stanley Trachtenberg, 147–51. Detroit, MI: Gale Research, 1982.

"Death Calls Ossian Guthrie." *Chicago Daily Tribune*, October 26, 1908.

Deloria, Philip J. *Playing Indian*. New Haven: Yale University Press, 2022.

Demuth, Bathsheba. *Floating Coast: An Environmental History of the Bering Strait*. New York: W. W. Norton, 2019.

———. "Turn and Live with Animals: The Slaughterhouse Ethic of Soviet and American Whalers Tells Us We Must Look beyond Communism and Capitalism to Survive." *Aeon*, October 29, 2019.

Denton, William. *Our Planet, Its Past and Future; or, Lectures on Geology*. Wellesley, MA: Denton, 1881.

Dimock, Wai-chee. *Empire for Liberty: Melville and the Poetics of Individualism*. Princeton, NJ: Princeton University Press, 1989.

Dyer, Michael P. *"O'er the Wide and Tractless Sea": Original Art of the Yankee Whale Hunt*. New Bedford, MA: Old Dartmouth Historical Society / New Bedford Whaling Museum, 2017.

Edwards, Mary K. Bercaw. *Cannibal Old Me: Spoken Sources in Melville's Early Works*. Kent, OH: Kent State University Press, 2009.

Estes, Nick. *Our History Is the Future: Standing Rock versus the Dakota Access Pipeline, and the Long Tradition of Indigenous Resistance*. New York: Verso, 2019.

Ewing, Heidi, and Rachel Grady. *Detropia*. DVD. Loki Films, 2013.

"Exit the Progress: Whaler Sunk at State Street Bridge by a Sand Scow." *Chicago Daily Tribune*, September 25, 1892.

Fassi, Anthony J. "Industrial Ruins, Urban Exploring, and the Postindustrial Picturesque." *CR: The New Centennial Review* 10, no. 1 (2010): 141–52.

"Fate of the Whaler Progress: The Famous Old Ship Becomes the Prey of Junk Dealers." *Chicago Daily Tribune*, July 22, 1895.

F. H. N. "Letter from Martha's Vineyard." *Christian Advocate and Journal*, August 17, 1865.

Fornoff, Carolyn, Patricia Eunji Kim, and Bethany Wiggin. "Coda." In *Timescales: Thinking across Ecological Temporalities*, edited by Carolyn Fornoff, Patricia Eunji Kim, and Bethany Wiggin, 193–201. Minneapolis: University of Minnesota Press, 2020.

Foster, Elizabeth S. "Another Note on Melville and Geology." *American Literature* 22, no. 4 (1951): 479–87.

———. "Melville and Geology." *American Literature* 17, no. 1 (1945): 50–65.

"Fred. Engelhart's Monster Whale." *Atchison Globe* (Kansas), July 10, 1883.

Freeburg, Christopher. *Melville and the Idea of Blackness: Race and Imperialism in Nineteenth-Century America*. New York: Cambridge University Press, 2012.

Frost, Robert. "A Masque of Mercy (1947)." In *The Poetry of Robert Frost*, edited by Edward Connery Lathem, 493–524. New York: Holt, Rinehart and Winston, 1969.

Frost, Samantha. *Biocultural Creatures: Toward a New Theory of the Human*. Durham, NC: Duke University Press, 2016.

Gassan, Richard H. *The Birth of American Tourism: New York, the Hudson Valley, and American Culture, 1790–1830*. Amherst: University of Massachusetts Press, 2008.

Gelles, David. "The Texas Group Waging a National Crusade against Climate Action." *New York Times*, December 4, 2022.

Ghosh, Amitav. *The Great Derangement: Climate Change and the Unthinkable*. Chicago: University of Chicago Press, 2016.

Gifford, Daniel. *The Last Voyage of the Whaling Bark* Progress: *New Bedford, Chicago and the Twilight of an Industry*. Jefferson, NC: McFarland, 2020.

Gilmore, David D. *Manhood in the Making: Cultural Concepts of Masculinity*. New Haven, CT: Yale University Press, 1990.

Godfrey, Edward K. *The Island of Nantucket: What It Was and What It Is. . . .* Boston: Lee and Shepard, 1882.

Gómez-Barris, Macarena. *The Extractive Zone: Social Ecologies and Decolonial Perspectives*. Durham, NC: Duke University Press, 2017.

Gordon, Robert J. *The Rise and Fall of American Growth: The U.S. Standard of Living since the Civil War*. Princeton, NJ: Princeton University Press, 2016.

A Guide to Martha's Vineyard and Nantucket, with a Directory of the Cottage City. Boston: Rockwell and Churchill, 1878.

Gumbs, Alexis Pauline. *Undrowned: Black Feminist Lessons from Marine Mammals*. Chico, CA: AK Press, 2020.

Harris, Andrew C., Paul R. Pentel, and Mark G. LeSage. "Prevalence, Magnitude, and Correlates of an Extinction Burst in Drug-Seeking Behavior in Rats Trained to Self-Administer Nicotine during Unlimited Access (23 h/d) Sessions." *Psychopharmacology*, no. 194 (2007): 395–402.

Harris, Daniel. *Cute, Quaint, Hungry and Romantic: The Aesthetics of Consumerism*. New York: Basic Books, 2000.

Harvey, Elizabeth D. "The Portal of Touch." *American Historical Review* 116, no. 2 (2011): 385–400.

Haynes, Melissa. "Animal." In *Fueling Culture: 101 Words for Energy and Environment*, edited by Imre Szeman, Jennifer Wenzel, and Patricia Yaeger, 35–38. New York: Fordham University Press, 2017.

Heath, Wm. Heath. "Whalers to Weavers: New Bedford's Urban Transformation and Contested Identities." *Journal for the Society for Industrial Archaeology* 40, no. 1–2 (2014): 7–32.

Henning, Michelle. "New Lamps for Old: Photography, Obsolescence, and Social Change." In *Residual Media*, edited by Charles R. Acland, 48–65. Minneapolis: University of Minnesota Press, 2007.

Hillway, Tyrus. "Melville's Geological Knowledge." *American Literature* 21, no. 2 (1949): 232–37.

Hohman, Elmo Paul. *The American Whaleman: A Study of Life and Labor in the Whaling Industry*. New York: Longmans, Green, 1928.

Hubbard, Elbert. *Little Journeys to the Homes of Great Business Men*. East Aurora, NY: Roycrofters, 1909.

Hubbert, M. King. "Nuclear Energy and the Fossil Fuels." Paper presented at the Spring Meeting of the Southern District, Division of Petroleum, American Petroleum Institute, San Antonio, TX, March 7, 1956.

Huhtamo, Erkki. "Dismantling the Fairy Engine: Media Archaeology as Topos Study." In *Media Archaeology: Approaches, Applications, and Implications*, edited by Erkki Huhtamo and Jussi Parikka, 27–47. Berkeley: University of California Press, 2011.

———. "(Un)walking at the Fair: About Mobile Visualities at the Paris Universal Exposition of 1900." *Journal of Visual Culture* 12, no. 1 (2013): 61–88.

Huhtamo, Erkki, and Jussi Parikka, eds. *Media Archaeology: Approaches, Applications, and Implications*. Berkeley: University of California Press, 2011.

Hurley, Jessica. *Infrastructures of Apocalypse: American Literature and the Nuclear Complex*. Minneapolis: University of Minnesota Press, 2020.

Hurley, Jessica, and Jeffrey Insko. "Introduction: The Infrastructure of Emergency." *American Literature* 93, no. 3 (2021): 345–59.

Ibrahim, Habiba. *Black Age: Oceanic Lifespans and the Time of Black Life*. New York: New York University Press, 2021.

Industrial Chicago. Vol. 4: *The Commercial Interests*. Chicago: Goodspeed, 1894.

Ingraham, Christopher. "The Entire Coal Industry Employs Fewer People Than Arby's." *Washington Post*, March 31, 2017.

Insko, Jeffrey. "'All of Us Are Ahabs': 'Moby-Dick' in Contemporary Public Discourse." *Journal of the Midwest Modern Language Association* 40, no. 2 (2007): 19–37.

———. "Extraction." In *The Cambridge Companion to Environmental Humanities*, edited by Jeffrey Jerome Cohen and Stephanie Foote. Cambridge: Cambridge University Press, 2021.

———. "On Dismantling: A Report from Michigan." *Resilience: A Journal of the Environmental Humanities* 8, no. 1 (2020): 139–53.

Jackson, Steven J. "Rethinking Repair." In *Media Technologies: Essays on Communication, Materiality and Society*, edited by Tarleton Gillespie, Kirsten A. Foot, and Pablo J. Boczkowski, 221–39. Cambridge, MA: MIT Press, 2014.

James, C. L. R. *Mariners, Renegades, and Castaways: The Story of Herman Melville and the World We Live In*. New York: C. L. R. James, 1953.

Jay, Martin. "In the Realm of the Senses: An Introduction." *American Historical Review* 116, no. 2 (2011): 307–15.

Jenner, Mark S. R. "Follow Your Nose? Smell, Smelling, and Their Histories." *American Historical Review* 116, no. 2 (2011): 335–51.

Johnson, Bob. *Carbon Nation: Fossil Fuels in the Making of American Culture.* Lawrence: University Press of Kansas, 2014.

———. *Mineral Rites: An Archaeology of the Fossil Economy.* Baltimore: Johns Hopkins University Press, 2019.

Jones, Christopher F. "Petromyopia: Oil and the Energy Humanities." *Humanities* 5, no. 2 (2016): 36–45.

———. *Routes of Power: Energy and Modern America.* Cambridge, MA: Harvard University Press, 2016.

Jones, Dan Burne. *The Prints of Rockwell Kent: A Catalogue Raisonné.* Chicago: University of Chicago Press, 1975.

Jones, Jamie L. "Fish out of Water: The 'Prince of Whales' Sideshow and the Environmental Humanities." *Configurations* 25, no. 2 (2017): 189–214.

———. "The Navy's Stone Fleet." *The Opinionator* (blog), *New York Times*, January 26, 2012. https://opinionator.blogs.nytimes.com/2012/01/26/the-navys-stone-fleet/.

———. "Oil: Viscous Time in the Anthropocene." *Los Angeles Review of Books*, March 22, 2016.

———. "Print Nostalgia: Skeuomorphism and Rockwell Kent's Woodblock Style." *American Art* 31, no. 3 (2017): 2–25.

Jonik, Michael. *Herman Melville and the Politics of the Inhuman.* Cambridge: Cambridge University Press, 2018.

Jørgensen, Dolly. "Naming and Claiming the Last." *The Return of Native Nordic Fauna* (blog), April 8, 2013. https://dolly.jorgensenweb.net/nordicnature/?p=450.

Jue, Melody. *Wild Blue Media: Thinking through Seawater.* Durham, NC: Duke University Press Books, 2020.

Kahn, Douglas. *Earth Sound Earth Signal: Energies and Earth Magnitude in the Arts.* Berkeley: University of California Press, 2013.

Kane, Joan Naviyuk. "Citation in the Wake of Melville." In *A Few Lines in the Manifest*, 25–43. Philadelphia: Albion Books, 2018.

Kent, Rockwell. "How I Make a Wood Cut." In *Enjoy Your Museum Series*, edited by Carl Thurson. Pasadena, CA: Esto, 1934.

———. *It's Me O Lord: The Autobiography of Rockwell Kent.* New York: Dodd, Mead, 1955.

———. *N by E.* New York: Random House, 1930.

———. *This Is My Own.* New York: Duell, Sloan and Pearce, 1940.

———. "Workers of the World, Unite!" *New Masses*, July 20, 1937, cover illustration.

King, Richard J. *Ahab's Rolling Sea: A Natural History of Moby-Dick.* Chicago: University of Chicago Press, 2019.

King, Tiffany Lethabo. *The Black Shoals: Offshore Formations of Black and Native Studies.* Durham, NC: Duke University Press, 2019.

Knappett, Carl. "Photographs, Skeuomorphs and Marionettes: Some Thoughts on Mind, Agency and Object." *Journal of Material Culture* 7, no. 1 (2002): 97–117.

LaDuke, Winona, and Deborah Cowen. "Beyond Wiindigo Infrastructure." *South Atlantic Quarterly* 119, no. 2 (2020): 243–68.

"Lake Shipping News." *Chicago Daily Tribune*, July 22, 1892.

Larkin, Brian. "The Politics and Poetics of Infrastructure." *Annual Review of Anthropology* 42, no. 1 (2013): 327–43.

Lauter, Paul. "Melville Climbs the Canon." *American Literature* 66, no. 1 (1994): 1–24.

Lavery, Grace E. *Quaint, Exquisite: Victorian Aesthetics and the Idea of Japan.* Princeton, NJ: Princeton University Press, 2019.

Lawrence, D. H. *Studies in Classic American Literature.* 1923; repr. London: Martin Secker, 1933.

Lawrence, Matthew. Interview by Jamie L. Jones, June 21, 2014.

Lears, T. J. Jackson. *No Place of Grace: Antimodernism and the Transformation of American Culture, 1880–1920.* New York: Pantheon Books, 1981.

Leary, John Patrick. "Detroitism." *Guernica*, January 15, 2011. www.guernicamag.com/leary_1_15_11/.

———. *Keywords for the Age of Austerity: A Lexicon of Inequality.* Chicago: Haymarket Books, 2018.

Leavitt, John F. *The Charles W. Morgan.* Mystic, CT: Mystic Seaport Museum, 1998.

Legge, Christopher. "Last Voyage of the Progress." *Bulletin of the Field Museum of Natural History* 40, no. 12 (1969): 5–6.

Leighton, Clare. *Wood Engraving of the 1930's.* London: Studio, 1936.

LeMenager, Stephanie. "Infrastructure Again, and Always." *Reviews in Cultural Theory* 6, no. 3 (2016): 25–29.

———. *Living Oil: Petroleum Culture in the American Century.* New York: Oxford University Press, 2014.

Lennon, Anastasia E. "America's Leader in Offshore Wind: What Vineyard Wind Final Approval Means for New Bedford." *Standard-Times* (New Bedford, MA), May 12, 2021.

Lerman, Dorothea E., and Brian A. Iwata. "Prevalence of the Extinction Burst and Its Attenuation during Treatment." *Journal of Applied Behavior Analysis* 28, no. 1 (1995): 93–94.

Lindgren, James M. "'Let Us Idealize Old Types of Manhood': The New Bedford Whaling Museum, 1903–1941." *New England Quarterly* 72, no. 2 (June 1999): 163–206.

Linthicum, Kent, Mikaela Relford, and Julia C. Johnson. "Defining Energy in Nineteenth-Century Native American Literature." *Environmental Humanities* 13, no. 2 (2021): 372–90.

Lipton, Eric. "'The Coal Industry Is Back,' Trump Proclaimed. It Wasn't." *New York Times*, October 5, 2020.

Lowell, Robert. *Collected Poems.* Edited by Frank Bidart and David Gewanter. New York: Farrar, Straus and Giroux, 2003.

Lowenthal, David. *The Heritage Crusade and the Spoils of History.* Cambridge: Cambridge University Press, 1998.

Luciano, Dana. "Sacred Theories of Earth: Matters of Spirit in William and Elizabeth Denton's *The Soul of Things.*" *American Literature* 86, no. 4 (2014): 713–36.

———. "Speaking Substances: Rock." *Los Angeles Review of Books*, April 12, 2016.

MacCannell, Dean. "Sightseeing and Social Structure: The Moral Integration of Modernity." In *Tourists and Tourism: A Reader*, 2nd ed., edited by Sharon Gmelch, 57–72. Long Grove, IL: Waveland, 2010.

Macekura, Stephen J. *The Mismeasure of Progress: Economic Growth and Its Critics.* Chicago: University of Chicago Press, 2020.

Madrigal, Alexis. *Containers*, 9 episodes. Produced by Fusion, podcast. https://podcasts.apple.com/us/podcast/containers/id1209559177.

Malm, Andreas. *Fossil Capital: The Rise of Steam Power and the Roots of Global Warming.* London: Verso, 2016.

Malthus, Thomas. *An Essay on the Principle of Population and Other Writings.* London: Penguin Classics, 2015.

"Mammoth Proportions." *Daily Inter Ocean* (Chicago), December 25, 1880.

Map of Nantucket: The Old Colony Line to Oak Bluffs, Martha's Vineyard, and Nantucket via New Bedford and Woods Holl. Old Colony Railroad, 1879.

Marchand, Yves, Romain Meffre, and Robert Polidori. *The Ruins of Detroit.* 5th ed. Göttingen: Steidl, 2013.

Martin, Constance. *Distant Shores: The Odyssey of Rockwell Kent.* Berkeley: University of California Press, 2000.

Martingale, Hawser [John Sherburne Sleeper]. *Salt Water Bubbles; or, Life on the Wave.* Boston: Wm. J. Reynolds, 1854.

Mathews, John Joseph. *The Osages: Children of the Middle Waters.* Norman: University of Oklahoma Press, 1961.

———. *Sundown.* Norman: University of Oklahoma Press, 1988.

Matthiessen, F. O. *American Renaissance: Art and Expression in the Age of Emerson and Whitman.* Oxford: Oxford University Press, 1941.

McCarron, Heather. "Offshore Wind Company Lays Final Cable at Barnstable Beach. What Is Next?" *Cape Cod Times*, January 19, 2023.

McClintock, Anne. "Monster: A Fugue in Fire and Ice." In *Oceans in Transformation*, edited by Nick Axel et al., e-flux Architecture and TBA21-Academy, 2020. www.e-flux.com/architecture/oceans/331865/monster-a-fugue-in-fire-and-ice/.

———. "Slow Violence and the BP Oil Crisis in the Gulf of Mexico: Militarizing Environmental Catastrophe." *Emisférica* 9, no. 1–2 (2012). https://hemisphericinstitute.org/en/emisferica-91/9-1-dossier/slow-violence-and-the-bp-oil-crisis-in-the-gulf-of-mexico-militarizing-environmental-catastrophe.html.

McCraw, Thomas K. *Prophet of Innovation: Joseph Schumpeter and Creative Destruction.* Cambridge, MA: Belknap Press of Harvard University Press, 2007.

McFee, William. "Foreword." In *Greasy Luck: A Whaling Sketchbook*, by Gordon Grant. 1932; repr. Mineola, NY: Dover, 2004.

McLuhan, Marshall. *Understanding Media: The Extensions of Man.* Cambridge, MA: MIT Press, 1994.

Meadows, Donella H., and Club of Rome, eds. *The Limits to Growth: A Report for the Club of Rome's Project on the Predicament of Mankind.* New York: Universe Books, 1972.

Melville, Herman. *Moby-Dick; or, The Whale.* Edited by Luther S. Mansfield and Howard P. Vincent. New York: Hendricks House, 1952.

Melville, Herman, and Dan McCall. "Benito Cereno." In *Melville's Short Novels: Authoritative Texts, Contexts, Criticism*, 34–102. New York: W. W. Norton, 2002.

Melville, Herman, and Hershel Parker. *Moby-Dick: An Authoritative Text, Contexts, Criticism.* 3rd ed. New York: W. W. Norton, 2018.

"A Memorial to Whaling: Colonel Green's Failure to Endow Museum Raises Problems of Financing." *New York Times*, June 6, 1937.

Miles, Tiya. "Nantucket Doesn't Belong to the Preppies." *Atlantic Monthly*, August 30, 2021. www.theatlantic.com/ideas/archive/2021/08/nantucket-doesnt-belong-to-the-preppies/619874/.

Miller, Angela. "Reading Ahab: Rockwell Kent, Herman Melville, and C. L. R. James." In *Renew Marxist Art History*, edited by Warren Carter, Barnaby Haran, and Frederic J. Schwartz, 223–45. London: Art/Books, 2013.

Miller, Elizabeth Carolyn. *Extraction Ecologies and the Literature of the Long Exhaustion*. Princeton, NJ: Princeton University Press, 2021.

Mitchell, Jon. "Remarks." Presented at the New Bedford Homecoming of the *Charles W. Morgan*, New Bedford, MA, June 28, 2014.

Mitchell, Timothy. *Carbon Democracy: Political Power in the Age of Oil*. London: Verso, 2013.

———. "Economentality: How the Future Entered Government." *Critical Inquiry* 40, no. 4 (2014): 479–507.

"The Monster Whale." *Evening Star* (Washington, DC), May 27, 1882.

Moore, Andrew, and Philip Levine. *Detroit Disassembled*. Akron, OH: Akron Art Museum, 2010.

Morison, Samuel Eliot. *The Maritime History of Massachusetts, 1783–1860*. Boston: Houghton Mifflin, 1921.

Morris, Edmund. *Derrick and Drill; or, An Insight into the Discovery, Development, and Present Condition and Future Prospects of Petroleum, in New York, Pennsylvania, West Virginia, &c.* New York: James Miller, 1865.

Mulligan, Frank. "New Bedford Offshore Wind Business Blows Its Stack." *Standard-Times* (New Bedford, MA), January 27, 2023.

Mumford, Lewis. *Herman Melville*. New York: Literary Guild of America, 1929.

———. *Technics and Civilization*. Chicago: University of Chicago Press, 2010.

Murison, Justine S. "'The Paradise of Non-Experts': The Neuroscientific Turn of the 1840s United States." In *The Neuroscientific Turn*, edited by Melissa M. Littlefield and Jenell Johnson, 29–48. Ann Arbor: University of Michigan Press, 2012.

———. *The Politics of Anxiety in Nineteenth-Century American Literature*. Cambridge: Cambridge University Press, 2011.

Nance, Susan. *Entertaining Elephants: Animal Agency and the Business of the American Circus*. Baltimore: Johns Hopkins University Press, 2013.

"Nautical Notes." *Daily Picayune* (New Orleans, LA), June 10, 1892.

Neely, Michelle C. *Against Sustainability: Reading Nineteenth-Century America in the Age of Climate Crisis*. New York: Fordham University Press, 2020.

Neurath, Paul. *From Malthus to the Club of Rome and Back: Problems of Limits to Growth, Population Control, and Migrations*. Armonk, NY: M. E. Sharpe, 1994.

Ngai, Sianne. *Our Aesthetic Categories: Zany, Cute, Interesting*. Cambridge, MA: Harvard University Press, 2012.

Nicholas, Tom. *VC: An American History*. Cambridge, MA: Harvard University Press, 2019.

Nixon, Rob. *Slow Violence and the Environmentalism of the Poor*. Cambridge, MA: Harvard University Press, 2013.

Norling, Lisa. *Captain Ahab Had a Wife: New England Women and the Whalefishery, 1720–1870*. Chapel Hill: University of North Carolina Press, 2000.

Norris, Frank. *The Octopus*. New York: Doubleday, Page, 1903.

Northrup, A. Judd. *'Sconset Cottage Life: A Summer on Nantucket Island*. New York: Baker, Pratt, 1881.

NPR Staff. "Transcript: Greta Thunberg's Speech at the U.N. Climate Action Summit." NPR.org, September 23, 2019. www.npr.org/2019/09/23/763452863/transcript-greta-thunbergs-speech-at-the-u-n-climate-action-summit.

O'Brien, Jean M. *Firsting and Lasting: Writing Indians out of Existence in New England*. Minneapolis: University of Minnesota Press, 2010.

Odell, Jenny. *How to Do Nothing: Resisting the Attention Economy*. Brooklyn, NY: Melville House, 2019.

O'Hara, Dan. "Skeuomorphology and Quotation." *Morphomata* 2 (2012): 281–94.

Old Places and New People; or, Our Pilgrimage, and What We Saw. New York: Leve and Alden's, 1881.

The Old Colony; or, Pilgrim Land, Past and Present. Fall River Line and Old Colony Railroad, 1887.

"Old Whaler Dedicated as Memorial to Industry." *Boston Daily Globe*, July 22, 1926.

Olson, Charles, Donald Allen, Benjamin Friedlander, and Robert Creeley. *Collected Prose*. Berkeley: University of California Press, 1997.

"The Only Manager Who Has Particularly Rejoiced." *Daily Inter Ocean* (Chicago), January 25, 1881.

Orlando, Francesco. *Obsolete Objects in the Literary Imagination: Ruins, Relics, Rarities, Rubbish, Uninhabited Places, and Hidden Treasures*. New Haven, CT: Yale University Press, 2006.

Otter, Samuel. *Melville's Anatomies*. Berkeley: University of California Press, 1999.

Parikka, Jussi. *A Geology of Media*. Minneapolis: University of Minnesota Press, 2015.

———. *Insect Media: An Archaeology of Animals and Technology*. Minneapolis: University of Minnesota Press, 2010.

———. *What Is Media Archaeology?* Cambridge, MA: Polity Press, 2012.

Parks, Lisa, and Nicole Starosielski. "Introduction." In *Signal Traffic: Critical Studies of Media Infrastructures*, ed. Lisa Parks and Nicole Starosielski. Urbana: University of Illinois Press, 2015.

Parsons, Amy. "'A Careful Disorderliness': Transnational Labors in Melville's *Moby-Dick*." *ESQ: A Journal of the American Renaissance* 58, no. 1 (2012): 71–101.

Peters, John Durham. "Infrastructuralism: Media as Traffic between Nature and Culture." In *Traffic: Media as Infrastructures and Cultural Practices*, edited by Marion Näser-Lather and Christoph Neubert, 29–49. Leiden: Brill, 2015.

———. *The Marvelous Clouds: Toward a Philosophy of Elemental Media*. Chicago: University of Chicago Press, 2016.

Philbrick, Thomas. *James Fenimore Cooper and the Development of American Sea Fiction*. Cambridge, MA: Harvard University Press, 2014.

Plumer, Brad, and Jim Tankersley. "Trump Picks Economic Winners, Guided by Nostalgia." *New York Times*, June 18, 2018.

The Popular Resorts and Fashionable Watering Places of Southeastern Massachusetts and Newport, R.I. Boston: Old Colony Railroad, 1878.

Powell, Miles A. *Vanishing America: Species Extinction, Racial Peril, and the Origins of Conservation.* Cambridge, MA: Harvard University Press, 2016.

The Presentation of the Whaleman Statue to the City of New Bedford by William W. Crapo and the Exercises at the Dedication, June Twentieth, Nineteen Hundred and Thirteen. New Bedford, MA: E. Anthony and Sons, 1913.

"Railrider Still Raiding Ohio." *Kalamazoo (MI) Daily Telegraph,* May 17, 1881.

Reinarz, Jonathan. *Past Scents: Historical Perspectives on Smell.* Urbana: University of Illinois Press, 2014.

Renker, Elizabeth. *The Origins of American Literature Studies: An Institutional History.* Cambridge: Cambridge University Press, 2007.

Riddell, John [Corey Ford]. *In the Worst Possible Taste.* New York: Charles Scribner's Sons, 1932.

Rifkin, Mark. *Beyond Settler Time: Temporal Sovereignty and Indigenous Self-Determination.* Durham, NC: Duke University Press, 2017.

———. "The Duration of the Land: The Queerness of Spacetime in Sundown." *Studies in American Indian Literatures* 27, no. 1 (2015): 33–69.

"'Right' Here: That Whale Is in the City, but Its Resting Place Is Concealed." *Daily Inter Ocean* (Chicago), December 24, 1880.

Robinson, Ione. *A Wall to Paint On.* New York: E. P. Dutton, 1946.

Roman, Joe. *Whale.* London: Reaktion, 2006.

Rosaldo, Renato. "Imperialist Nostalgia." *Representations,* no. 26 (1989): 107–22.

Rozwadowski, Helen M. *Vast Expanses: A History of the Oceans.* London: Reaktion Books, 2018.

Rubenstein, Michael, Bruce Robbins, and Sophia Beal. "Infrastructuralism: An Introduction." *MFS Modern Fiction Studies* 61, no. 4 (2015): 575–86.

Rudwick, M. J. S. *Worlds before Adam: The Reconstruction of Geohistory in the Age of Reform.* Chicago: University of Chicago Press, 2008.

Ruskin, John. *Ariadne Florentina: Six Lectures on Wood and Metal Engraving, Given before the University of Oxford, in Michaelmas Term, 1872.* Sunnyside, UK: George Allen, 1876.

Rydell, Robert W. *All the World's a Fair: Visions of Empire at American International Expositions, 1876–1916.* Chicago: University of Chicago Press, 1987.

Schell, Jennifer. *A Bold and Hardy Race of Men: The Lives and Literature of American Whalemen.* Amherst: University of Massachusetts Press, 2013.

Schmelzer, Matthias. *The Hegemony of Growth: The OECD and the Making of the Economic Growth Paradigm.* Cambridge: Cambridge University Press, 2016.

Schmelzer, Matthias, Andrea Vetter, and Aaron Vansintjan, *The Future Is Degrowth: A Guide to a World beyond Capitalism.* London: Verso, 2022.

Schneider-Mayerson, Matthew. *Peak Oil: Apocalyptic Environmentalism and Libertarian Political Culture.* Chicago: University of Chicago Press, 2015.

Schooler, Lynn. *The Last Shot: The Incredible Story of the CSS* Shenandoah *and the True Conclusion of the American Civil War.* New York: Ecco, 2005.

Schrey, Dominik. "Analogue Nostalgia and the Aesthetics of Digital Remediation." In *Media and Nostalgia: Yearning for the Past, Present and Future*, edited by Katharina Niemeyer, 27–38. New York: Palgrave MacMillan, 2014.

Schuller, Kyla. "Speaking Substances: Bodies." *Los Angeles Review of Books*, March 23, 2016.

Schultz, Elizabeth A. "Melville's Environmental Vision in Moby-Dick." *Interdisciplinary Studies in Literature and Environment* 7, no. 1 (2000): 97–113.

———. *Unpainted to the Last:* Moby-Dick *and Twentieth-Century American Art*. Lawrence: University Press of Kansas, 1995.

Schumpeter, Joseph A. *Capitalism, Socialism, and Democracy*. New York: Harper Perennial, 1942.

Scott, Heidi. "Whale Oil Culture, Consumerism, and Modern Conservation." In *Oil Culture*, edited by Ross Barrett and Daniel Worden, 3–18. Minneapolis: University of Minnesota Press, 2014.

Segal, David. "In Brexit Vote, Town's Nostalgia for Seafaring Past Muddied Its Future." *New York Times*, April 23, 2018.

Sekula, Allan. *Fish Story*. Rotterdam: Witte de With Center for Contemporary Art, 1995.

Sekula, Allan, and Noël Burch. "Notes for a Film." In *The Forgotten Sea*. Icarus Films Home Video, 2015.

Sekula, Allan, Laleh Khalili, and B. H. D. Buchloh. *Fish Story*. 3rd rev. English ed. London: MACK, 2018.

Shannon, Laurie. "Tallow." In *Fueling Culture: 101 Words for Energy and Environment*, edited by Imre Szeman, Jennifer Wenzel, and Patricia Yaeger, 346–48. New York: Fordham University Press, 2017.

Sharpe, Christina Elizabeth. *In the Wake: On Blackness and Being*. Durham, NC: Duke University Press, 2016.

Shoemaker, Nancy, ed. *Living with Whales: Documents and Oral Histories of Native New England Whaling History*. Amherst: University of Massachusetts Press, 2014.

———. *Native American Whalemen and the World: Indigenous Encounters and the Contingency of Race*. Chapel Hill: University of North Carolina Press, 2015.

Shukin, Nicole. *Animal Capital: Rendering Life in Biopolitical Times*. Minneapolis: University of Minnesota Press, 2009.

Shumway, David R. *Creating American Civilization: A Genealogy of American Literature as an Academic Discipline*. Minneapolis: University of Minnesota Press, 1994.

Simpson, Mark, and Imre Szeman. "Impasse Time." *South Atlantic Quarterly* 120, no. 1 (2021): 77–89.

Simpson, Audra. *Mohawk Interruptus: Political Life across the Borders of Settler States*. Durham, NC: Duke University Press, 2014.

Skinner, B. F. *The Behavior of Organisms: An Experimental Analysis*. New York: Appleton-Century-Crofts, 1938.

Sloan, John. *New York Scene: 1906–1913*. London: Routledge, 2013.

Slotkin, Richard. *Regeneration through Violence: The Mythology of the American Frontier, 1600–1860*. Norman: University of Oklahoma Press, 2000.

Smith, Jason F. "Thou Uncracked Keel: The Many Voyages of the Whaleship Charles W. Morgan and the Presence of the American Maritime Past." *New England Quarterly* 89, no. 3 (2016): 421–56.

Smithson, Robert, and Jack D. Flam. "Entropy and the New Monuments" [1966]. In *Robert Smithson, the Collected Writings*, 10–23. Berkeley: University of California Press, 1996.

Spanos, William V. *The Errant Art of* Moby-Dick: *The Canon, the Cold War, and the Struggle for American Studies*. Durham, NC: Duke University Press, 1995.

Spark, Clare. *Hunting Captain Ahab: Psychological Warfare and the Melville Revival*. Kent, OH: Kent State University Press, 2001.

Spiro, Jonathan Peter. *Defending the Master Race: Conservation, Eugenics, and the Legacy of Madison Grant*. Hanover, NH: University Press of New England, 2009.

Star, Susan Leigh. "The Ethnography of Infrastructure." *American Behavioral Scientist* 43, no. 3 (1999): 377–91.

Starosielski, Nicole. "Beyond the Sun: Embedded Solarities and Agricultural Practice." *South Atlantic Quarterly* 120, no. 1 (2021): 13–24.

———. *The Undersea Network*. Durham, NC: Duke University Press, 2015.

Starosielski, Nicole, Braxton Soderman, and Cris Cheek. "Introduction." *Amodern* 2, no. Network Archaeology (October 2013). http://amodern.net/article/network-archaeology/.

Sterne, Jonathan. "Out with the Trash: On the Future of New Media." In *Residual Media*, edited by Charles R. Acland, 16–31. Minneapolis: University of Minnesota Press, 2007.

Stevens, Mrs. Mark. *Six Months at the World's Fair*. Detroit, MI: Detroit Free Press, 1895.

Stories about the Whale; with an Account of the Whale Fishery, and of the Perils Attending Its Prosecution. Concord, NH: Rufus Merrill, 1853.

Szeman, Imre. *After Oil*. Morgantown: West Virginia University Press, 2016.

———. "System Failure: Oil, Futurity, and the Anticipation of Disaster." *South Atlantic Quarterly* 106, no. 4 (2007): 805–23.

Szeman, Imre, and Darin Barney. "Introduction." *South Atlantic Quarterly* 120, no. 1 (2021): 1–11.

Szeman, Imre, and Dominic Boyer, eds. *Energy Humanities: An Anthology*. Baltimore: Johns Hopkins University Press, 2017.

———. "Introduction: On the Energy Humanities." In *Energy Humanities: An Anthology*, edited by Imre Szeman and Dominic Boyer, 1–13. Baltimore: Johns Hopkins University Press, 2017.

Szeman, Imre, Jennifer Wenzel, and Patricia Yaeger, eds. "Whaling." In *Fueling Culture: 101 Words for Energy and Environment*, 373–75. New York: Fordham University Press, 2017.

TallBear, Kim. "Badass Indigenous Women Caretake Relations: #standingrock, #idlenomore, #blacklives Matter." In *Standing with Standing Rock: Voices from the #NoDAPL Movement*, edited by Nick Estes and Jaskiran Dhillon, 13–18. University of Minnesota Press, 2019.

"Tax Hearing Turns on Green's Hobbies." *New York Times*, January 14, 1938.

Taylor, Dorceta E. *The Rise of the American Conservation Movement: Power, Privilege, and Environmental Protection*. Durham, NC: Duke University Press, 2016.

Taylor, V., and F. Trentmann. "Liquid Politics: Water and the Politics of Everyday Life in the Modern City." *Past and Present* 211, no. 1 (2011): 199–241.

Territorial Agency, John Palmesino, and Ann-Sofi Rönnskog. "Radical Conservation: The Museum of Oil." In *Reset Modernity!*, edited by Bruno Latour, Christophe Leclercq, and Zentrum für Kunst und Medientechnologie Karlsruhe. Cambridge, MA: MIT Press, 2016.

The Tip End of Yankee Land; or, The Log Book of a Trip to Cape Cod, "The Vineyard," and "Old Nantuck," by a Recent Traveler. Published for the Benefit of All Lovers of the Quaint, the Picturesque, and the Historic. Fall River and Newport Lines, and Old Colony Railroad, 1885.

Todd, Zoe. "Fish, Kin and Hope: Tending to Water Violations in Amiskwaciwâskahikan and Treaty Six Territory." *Afterall: A Journal of Art, Context and Enquiry* 43 (2017): 102–7.

Tompkins, Kyla Wazana. "Sweetness, Capacity, Energy." *American Quarterly* 71, no. 3 (2019): 849–56.

Tompkins, Lucy. "Sierra Club Says It Must Confront the Racism of John Muir." *New York Times*, July 22, 2020.

Tondre, Michael. "Conrad's Carbon Imaginary: Oil, Imperialism, and the Victorian Petro-Archive." *Victorian Literature and Culture* 48, no. 1 (2020): 57–90.

Tønnessen, J. N., and Arne Odd Johnsen. *The History of Modern Whaling*. Berkeley: University of California Press, 1982.

Tourists' Guide to Down the Harbor, Hull and Nantasket, Downer Landing, Hingham, Cohasset, Marshfield, Scituate, Duxbury, "The Famous Jerusalem Road," "Historic Plymouth," Cottage City, Martha's Vineyard, Nantucket and the Summer Resorts of Cape Cod and the South Shore of Massachusetts. Boston: John F. Murphy, 1890.

"The Traveled Whale." *Kalamazoo (MI) Daily Telegraph*, May 8, 1886.

Traxel, David. *An American Saga: The Life and Times of Rockwell Kent*. New York: Harper and Row, 1980.

Underwood, Ted. *The Work of the Sun: Literature, Science, and Political Economy, 1760–1860*. New York: Palgrave Macmillan, 2005.

Vanderbilt, Kermit. *American Literature and the Academy: The Roots, Growth, and Maturity of a Profession*. Philadelphia: University of Pennsylvania Press, 1986.

"Wanderer's Crew All Safe on Shore: Last Eight Land at Cuttyhunk." *Boston Daily Globe*, August 27, 1924.

Warren, Elizabeth. "Remarks." Presented at The New Bedford Homecoming of the *Charles W. Morgan*, New Bedford, MA, June 28, 2014.

Warrin, Donald. *So Ends This Day: The Portuguese in American Whaling, 1765–1927*. North Dartmouth, MA: University of Massachusetts Dartmouth, Center for Portuguese Studies and Culture, 2010.

"Waterways and Water Pageants at the Fair." *Chicago Daily Tribune*, September 17, 1893.

Weaver, Raymond M. *Herman Melville: Mariner and Mystic*. New York: George H. Doran, 1939.

Webster, Robert M., and Bruce Erickson. "The Last Word?" *Nature* 380, no. 6573 (1996): 386.

Weinberg, Jonathan. "I Want Muscle: Male Desire and the Image of the Worker in American Art of the 1930s." In *The Social and the Real: Political Art of the 1930s in the Western Hemisphere*, edited by Alejandro Anreus, Diana L. Linden, and Jonathan Weinberg, 115–34. University Park: Penn State University Press, 2006.

Wenzel, Jennifer. "Introduction." In *Fueling Culture: 101 Words for Energy and Environment*, edited by Imre Szeman, Jennifer Wenzel, and Patricia Yaeger, 1–16. New York: Fordham University Press, 2017.

———. "Petro-Magic-Realism: Toward a Political Ecology of Nigerian Literature." *Postcolonial Studies* 9, no. 4 (2006): 449–64.

———. "Petro-Magic-Realism: Toward a Political Ecology of Nigerian Literature." In *Energy Humanities: An Anthology*, edited by Imre Szeman and Dominic Boyer, 486–504. Baltimore: Johns Hopkins University Press, 2017.

———. "Petro-Magic-Realism Revisited: Unimagining and Reimagining the Niger Delta." In *Oil Culture*, edited by Ross Barrett and Daniel Worden, 211–25. Minneapolis: University of Minnesota Press, 2014.

West, Richard V. "Rockwell Kent: After the Odyssey." In Constance Martin, *Distant Shores: The Odyssey of Rockwell Kent*, 113–21. Berkeley: University of California Press, 2000.

Whalan, Mark. "'Oil Was Trumps': John Dos Passos' USA, World War I, and the Growth of the Petromodern State." *American Literary History* 29, no. 3 (2017): 474–98.

"The Whale: A Direct Descendent of Jonah's Friend at Twentieth and Market." *North American* (Philadelphia), March 29, 1881.

"Whaler Coming Back from Chicago." *New York Times*, 1896.

"Whaler Ditched." *Boston Daily Advertiser*, June 30, 1896.

"Whaler Preserved in Concrete Recalls New Bedford Romance." *Boston Daily Globe*, July 26, 1925.

"Whaler Progress Sunk." *Milwaukee Sentinel*, September 25, 1892.

"Whales at the Fair: To Be Sighted and Caught by Whalers of the Progress." *Chicago Daily Tribune*, July 24, 1892.

"A Whaling Big Boat." *Daily Inter Ocean* (Chicago), May 8, 1892.

"Whaling Exhibit Is Purchased: Addition Is Made to the Field Columbian Museum." *Chicago Daily Tribune*, February 21, 1894.

White, Richard. *Railroaded: The Transcontinentals and the Making of Modern America*. New York: W. W. Norton, 2011.

White, Stephen C. Interview by Jamie L. Jones, June 3, 2021.

Whyte, Kyle Powys. "The Dakota Access Pipeline, Environmental Injustice, and U.S. Colonialism." *Red Ink* 19, no. 1 (2017): 154–69.

———. "Indigenous Climate Change Studies: Indigenizing Futures, Decolonizing the Anthropocene." *English Language Notes* 55, no. 1–2 (2017): 153–62.

Wien, Jake Milgram. "Rockwell Kent." Review of *The Prints of Rockwell Kent: A Catalogue Raisonné*, by Dan Burne Jones and Robert Rightmire. *Print Quarterly* 20, no. 2 (2003): 203–8.

Williams, Nicole J. "'Whalecraft': Clifford Warren Ashley and Whaling Craft Culture in Industrial New Bedford." *Journal of Modern Craft* 11, no. 3 (2018): 185–217.

Williams, Raymond. *Marxism and Literature*. Oxford: Oxford University Press, 1977.
Wilson, Terry P. *The Underground Reservation: Osage Oil*. Lincoln: University of Nebraska Press, 1985.
"A Wonderful Lady Rider." *Daily Alta California* (San Francisco), January 1, 1884.
Wood, Molly, and Scott Tong. "Counting Up American Coal Jobs: What's the Real Total?" *Marketplace*, June 15, 2017. www.marketplace.org/2017/06/15/sustainability/counting-american-coal-jobs-whats-real-total.
Woodward Jr., Ralph. "Cruise of Wanderer May Be Last from This Port." *New Bedford (MA) Times*, August 16, 1924. New Bedford Whaling Museum Research Library.
Wyer, Henry S. *Nantucket Characters: Indelible Photographs*. New York: Albertype, 1892.
Yaeger, Patricia. "Editor's Column: Literature in the Ages of Wood, Tallow, Coal, Whale Oil, Gasoline, Atomic Power, and Other Energy Sources." *PMLA* 126, no. 2 (2011): 305–26.
Yergin, Daniel. "Ensuring Energy Security." *Foreign Affairs* 85, no. 2 (2006): 69.
Zallen, Jeremy. *American Lucifers: The Dark History of Artificial Light, 1750–1865*. Chapel Hill: University of North Carolina Press, 2019.

INDEX

Page numbers in italics refer to illustrations.

Acushnet (ship), 55
aesthetics, 70–78, 174
Ahab (character), *160*, 202n87. See also *Moby-Dick* (Melville)
Alaimo, Stacy, 101, 102
ambergris, 4
American Car and Foundry Company, 175–79, *177*, 180
American Nervousness (Beard), 82
American Renaissance (Matthiessen), 150
American Whaleman, The (Hohman), 123
Animal Capital (Shukin), 99
Anishinaabe, 40
Apple (company), 167–68
Aquinnah, 188
Aron, Cindy, 81, 83
Arts and Crafts movement, 179, 180, 183
Ashcan School, 174
Ashley, Clifford Warren, 117, 118, 131, 137, 141, 146; *The Sea Chest*, *138*; *The Yankee Whaler*, 131, 146

baleen, 4, 5, 22, 44, 101. *See also* whale, as animal
Barnard, John Levi, 45, 52, 121
Barnum, P. T., 96–97, 112
beachy dreams, 19–20
Beard, George M., 81–84, 85–86; *American Nervousness*, 82
Benito Cereno (Melville), 186
Bennett, Jane, 101
Benson, Richard, 159
Berardi, Franco, 83
Bering Strait history, 12
Biden administration, 191

Birth of a Nation, The (film), 135–36
Black communities, 91, 122–23, 127, 186, 216n3
Blair, Gordon S., 139–41
Bluemel, Kristin, 162
Blum, Hester, 19, 37, 91, 94, 99, 182, 206n7
Bolter, J. David, 95, 100
bone in her teeth, as phrase, 186, 194
Boym, Svetlana, 164
Boyton, Paul, 98, 207n23
British maritime fleet and fuel, 17
British whaling industry, 213n7
broken world thinking, 61, 184
Brown, Dona, 78
Browne, J. Ross, 141; *Etchings of a Whaling Cruise*, 141
Buell, Frederick, 11
Burch, Noël, 89; *The Forgotten Space*, 89, 91
Burma, 196n26
Burtynsky, Edward, 184
butchering. *See* slaughtering

Call Me Ishmael (Olson), 156–57
Cambridge History of American Literature (publication), 150
Cape Verdeans, 122–23, 127–28, 189
capitalism, 21–22, 29–35, 48–50, 56–58, 88, 179; and economic growth, 29–32, 44–45, 199n4
Casarino, Cesare, 58
Cassidy, Donna, 174
Charles Phelps (ship), 107–8. See also *Progress* (ship)
Charles W. Morgan (ship), 26, 142–47, 156, 183–91, 194, 212n71

{ 237 }

Cherokee, 42
Chicago, IL, 93, 94, 97, 98, 101, 102, 108–14
Chicago Architecture Biennial, 192
Chukchi, 6
Church, Albert Cook, 118, 141
"Citation in the Wake of Melville" (Kane), 43–44
City of Taunton (ship), 71, 72, 204n25
Civil War, US, 4–5
Clifton, Elmer, 135, 136–42
climate change, 13, 16–18, 26, 30, 120–21, 147–49, 186, 189–90
coal, 24, 47, 62, 120, 147–48. *See also* fossil fuels; railroad travel; steamship travel
Cohen, Margaret, 19, 87, 182; *The Novel and the Sea*, 19
Columbian Exposition (1893), 23, 93, 108–14, *110*, 211n59
commercial art, 170–73
communism, 168
Conn, Steven, 146
containerization, 17, 18–19, 20, 88–89, 92, 94
Corbin, Alain, 83, 87, 91
Coronil, Fernando, 33
corpse, as entertainment. *See* Prince of Whales; whale, as animal
Covarrubias, Miguel, 171; *Covarrubias*, *172*
COVID-19 pandemic, 186
Cowen, Deborah, 37, 40
Crapo, William, 118, 125–26, *127*, 143
creative destruction, as term, 9, 21, 33, 53, 58, 193
critical race studies, 19
Cronon, William, 93
cuteness, 73–74. *See also* quaintness
Cutler, Carl, 183
Cuvier, Georges, 46

Daggett, Cara New, 11, 22, 62–63, 151
Daggoo (character), 41, 126. See also *Moby-Dick* (Melville)
Davis, Edward Wyatt, 169–70

Deepwater Horizon blowout (2010), 18, 37, 192
Demuth, Bathsheba, 12, 131
Denton, William, 47–48
Detroit, Michigan, 64–65, 89, 97; Michigan Central Station in, 64–65
Dowden, James, 109, 112
Down to the Sea in Ships (film), 135, 136–42, *140*
Drake, Edwin, 134

economic growth and capitalism, 21–22, 29–35, 44–45, 48–50, 56–58, 179, 199n4
electrotype method, 180
Ellis, Charles, 2
endlings, 142–47, 212n60
energy, as concept, 2–3, 7–8, 11–12, 195n9, 213n6. *See also* Daggett, Cara New
energy archaeology, 14–17, 31, 153–57, 167
energy humanities, as discipline, 11, 15–17, 63, 154, 161, 196n21, 197n38, 197n40, 203n5
energy impasse, 13–14
energy transition, 3–5, 7–8, 12–14, 93–95, 151–53, 154–56, 161, 189–91
Engelhardt, Fred J., 97–98, 102, 104, 105–6
Enjoy Your Museum series, 168–69
entertainment. *See* whaling entertainment
environmental injustice, 42, 186
environmentalism, 128–34
Estes, Nick, 42
Etchings of a Whaling Cruise (Browne), 141
exhaustion, 80–86. *See also* quaintness
extinction: as concept, 120–21, 135; of extractive industries, 16, 148; of Indigenous peoples, supposed, 42, 120, 201n45, 210n8; mass planetary, 30, 103; sport hunting and, 128–29; of whales and whaling industry, 24, 25, 31, 45, 48, 133, 202n71; white

supremacy and anxiety about, 119, 124, 134, 142, 143, 147. *See also* lasting
extraction, 13, 18, 32, 37, 49–50, 200n17. *See also* economic growth and capitalism; fishing industry; fossil fuels; whaling industry
extractivist nostalgia, 119–20, 135, 147–48, 191. *See also* nostalgia

Fall River Line, 67, 68, 69–70
fertility, 56–57
Field Museum, 110
firsting, 120–21. *See also* lasting
Firsting and Lasting (O'Brien), 16, 55, 120–21. *See also* firsting; lasting
Fisher Body Plant 21, Detroit, 64
fishermen, 174–75
fishing industry, 189–90
Fish Story (Sekula), 18, 88–91
Ford, Corey, 170–71; "N by G," 170–71
forgetting of the sea, 18–20, 87–91. *See also* hydrophasia
Forgotten Space, The (Burch), 89, 91
fossil fuels: development of, 3–4, 195n11; dismantling of, 192–93; extractive violence of, 200n17; labor and production of, 7; sources of, 196n21; tourism and, 66–70. *See also* oil industry; petroleum
fossil modernity, xi, 7–10, 16–17, 181, 196n21
Foucault, Michel, 14
Frost, Samantha, 12

Gassan, Richard H., 66–67
Gifford, Daniel W., 108, 111, 112, 208n59, 209n73, 209n78
Godfrey, Edward K., 66, 75–76
Golden Day, The (Mumford), 155
"Grand Ball Given by the Whales in Honor of the Discovery of the Oil Wells in Pennsylvania" (*Vanity Fair*), 2, *3*, 8, 13
Grant, Madison, 129, 130; *The Passing of the Great Race*, 130

Greasy Luck (McFee), 132
Green, Edward Howland Robinson, 118, 144–46, 183
Green, Hetty, 118
Griffith, D. W., 135
grotesque humor, 103, 105–6
Grusin, Richard, 95, 100
Guthrie, Ossian, 108, 109

Harris, Daniel, 74
Hartley, Marsden, 174
Harvey, Elizabeth D., 101–2
Henri, Robert, 174
Herman Melville (Mumford), 150, 214n20
Herman Melville: Mariner and Mystic (Weaver), 141, 150, 153–54
Hohman, Elmo, 123; *The American Whaleman*, 123
Howland, Abraham, 2
How to Do Nothing (Odell), 193
Hubbert, M. King, 48. *See also* peak oil
Huhtamo, Erkki, 14, 167
hunting, 43–44, 45, 128–34, 211n31. *See also* extinction
hydrophasia, 19. *See also* forgetting of the sea
hypermediacy, 104

Icelandic whaling industry, 6
Indigenous peoples: Melville and *Moby-Dick* on, 40, 42–43, 201n48; oil extraction and, 41–42; whaling and, 6, 42–44, 188, 201n46
Indigenous studies: as discipline, 19; critiques of settler colonial infrastructure in, 16–17, 52–53, 41–42, 198n60, 200n34
infrastructure: metaphors of, 36–40, 204n14; ruins of, 203n94, 203n96, 204n8, 204n12, 204n29; speculative infrastructure, 191; whaling and, 4–5, 36–44, 111–13, 117–20, 203n94. *See also* railroad travel; ruin pornography; steamship travel

INDEX   { 239

infrastructure studies, 15, 95–96, 200n23, 198n56. *See also* media studies
Insko, Jeffrey, 26, 37, 192–93
International Whaling Commission (IWC), 6
In the Worst Possible Taste (collection), 170
Iñupiat, 6, 43–44
Ishmael (character), 29–31, 181. See also *Moby-Dick* (Melville)

Jackson, Steven, 61, 184
Japan, 6, 75
jawbone, 40–41. *See also* whale, as animal
Johnson, Bob, 14, 81
Jones, Christopher, 4, 17, 36–37, 94, 197n40
Jue, Melody, 95–96, 99

Kane, Joan Naviyuk, 43–44; "Citation in the Wake of Melville," 43–44
Kent, Rockwell, 25; commercial art enterprise of, 170–73; content of work by, 173–80, 213n8, 214n30; *Fire Wood, 166*; *Hail and Farewell, 178*; "How I Make a Wood Cut," 168–69; illustrated edition of *Moby-Dick* by, 151–53, *160*; *It's Me O Lord*, 174–75; *N by E*, 170, 173; *Night Watch*, 175, *177*; political art by, 162–63; *Toilers of the Sea*, 175, *176*; Viking Press, colophon for, *162*; *Voyaging Southward from the Strait of Magellan*, 173; *Wilderness*, 165, 166, 173; woodblock form of, 157–63; woodblock method of, 163–66, 179–80; *Workers of the World, Unite!*, 163
kinship, 16, 52–53
Ku Klux Klan, 135–36
Kunstler, James Howard, 49; *The World Made by Hand*, 49

labor and political art, 162–63, 165–66, 170, 174–75, 181–82

LaDuke, Winona, 37, 40
Lakeside Press, 25, 157
lasting, 120–21, 143. *See also* extinction
Lavery, Grace, 74–75
Lawrence, D. H., 156
Lawrence, Matthew, 194
Lears, T. J. Jackson, 179
Leary, John Patrick, 65
Leighton, Clare, 161, 164, 173
LeMenager, Stephanie, 3–4, 15, 99, 139
Lenin Peace Prize, 168
Leopold, Aldo, 211n31
livestock industry, 6, 96, 132–34
loomings, 21–22, 32
Lowell, Robert, 184, 216n3
Lowenthal, David, 123
Lower Brule Sioux, 42
Lyell, Charles, 46

Macomber, John Arnold, 2
Maine, 174–75, 180, 181, 216n70
Makah, 6
Malthus, Thomas, 30
manifest dismantling, 193
Marine Commerce Terminal, 190, 191
Maritime History of Massachusetts (Morison), 88
Martha's Vineyard, 43, 63, 68, 142, 185, 188, 194
Marx, Karl, 29
Marxism, 162–63, 168
Matthiessen, F. O., 150; *American Renaissance*, 150
Mayflower (ship), 114
McCarthy, Joseph, 168
McClintock, Anne, 18
McFee, William, 132, 211n38; *Greasy Luck*, 132
McKee, Raymond, 139
McLachlan, Richard Lewis, 128
McLean, Malcolm, 18
McLuhan, Marshall, 15, 99, 114; *Understanding Media*, 15, 114
media, as concept, 15, 95. *See also* media studies; whale media

media studies, 8–9, 14–15, 19, 33, 37, 95–96, 99–101, 114, 198n50, 198n56, 207n25
Melville, Herman, 16, 21, 186, 195n18; *Benito Cereno*, 186. See also *Moby-Dick* (Melville)
Melville Revival, 6, 24–25, 150–58, 180–82. See also *Moby-Dick* (Melville)
Merrill, Rufus, 41
Middle Passage, 19
Miles, Tiya, 91
Miller, Angela, 158
Miller, Elizabeth, 45–46
Mitchell, Jon, 190–91
Mitchell, Timothy, 30, 195n11
Moby-Dick (Melville): early petroleum culture and, 44–48; Egyptian mother and fertility, 56–57; as energy theory, 21, 29–58; infrastructural metaphors in, 36–40; Kent's illustrated edition of, 25, 157–66, *160*; Melville Revival and, 6, 24–25, 150–58; nostalgia and, 180–82; obsolescence and, 53–58, 65; whaleman statue and, *125*, 126; whaling industry in, 29–36. See also Melville, Herman; Melville Revival
modernity, as term, 196n26
Monhegan Island, 174, 180. See also Maine
Morgan. See *Charles W. Morgan* (ship)
Morison, Samuel Eliot, 88; *Maritime History of Massachusetts*, 88
Morris, Edmund, 47
Mumford, Lewis, 15, 150, 154–57, 161; *The Golden Day*, 155; *Herman Melville*, 150, 214n20; *Technics and Civilization*, 154, 155
Museum of Oil, 192
Mystic Seaport Museum, 26, 139, 183, 185, 188, 194

Nabokov, Vladimir, 58
Nantucket, 53–55, 61–66, 72–73, 75–80, 84–86, 181, 189, 193, 205n44

Nantucket Characters (Wyer), 78–80, *80*
Nantucket Steam Boat Company, 63
National Audubon Society, 129
"N by G" (Ford), 170–71
nervousness, 82–83, 205n51. See also exhaustion
network archaeology, 14. See also infrastructure studies; media studies
neurasthenia, 82
New Bedford community and whaling industry, 29, 33–35, 53, 61–62, 71–72, 123, 142–47, 188–91
New Bedford Public Library, 124, *125*
New Bedford Whaling Museum, 71, 189, 212n59, 212n71
New Masses (publication), 163
Newton, George, 22, 96–97, 104
Ngai, Sianne, 73, 74. See also cuteness
Nigeria, 33, 34
Nixon, Rob, 16
NOAA (National Oceanic and Atmospheric Administration), 188
Norris, Frank, 105; *The Octopus*, 105
Northrup, A. Judd, 76–77, 81, 84
North Saskatchewan River oil spill (2016), 52
Norwegian whaling industry, 6, 213n7
nostalgia: extractivist, 119–20, 135, 147–48, 191; in Kent's illustrations, 151, 152–53, *166*, 173–80, *181*; in Melville's *Moby-Dick*, 180–82; for whaling culture, 23–24, 119–20
Novel and the Sea, The (Cohen), 19
nudity in art, 175–79
Nuu-chah-nulth, 6

O'Brien, Jean, 55, 120; *Firsting and Lasting*, 16, 55, 120–21. See also firsting; lasting
obsolescence, 9, 53–58, 194
Ocean House Hotel, Nantucket, 63
oceanic energy, 17–20
oceanic studies, 19, 206n9
Octopus, The (Norris), 105

INDEX { 241

Odell, Jenny, 193; *How to Do Nothing*, 193
offshore wind energy, 189–90, 191
O'Hara, Dan, 167
oil industry: labor and production in, 2; modern life and, 192, 195n10, 196n21; in Nigeria, 33, 34; in Pennsylvania, 1. *See also* fossil fuels; petroleum
oil media, 15, 99, 207n26
oil spills, 18, 37, 52
Old Colony, The (publication), 85
Old Colony Railroad, 63, 67–69
Old Dartmouth Historical Society, 118, 125, 128, 141, 212n71
Olson, Charles, 156–57, 214n25; *Call Me Ishmael*, 156–57
Osage, 42

Palmesino, John, 192
Parikka, Jussi, 14
Parks, Lisa, 95
Parsons, Amy, 182
Passing of the Great Race, The (Grant), 130
Peabody Museum, 110
peak oil, 48–53. *See also* oil industry
Pennsylvania oil boom, 1
Pequod (ship), 49, 54–55, 203n93. *See also Moby-Dick* (Melville)
Pequod, as term, 42
Permanent Subcommittee on Investigations, 168
Peters, John Durham, 15
petroleum, 1, 44–48. *See also* fossil fuels; oil industry
petro-magic, 33–35
petromodernity, as term, 15, 196n21
petromyopia, 17
pig-sticking, 132–33
Pinchot, Gifford, 129
Pioneer Inland Whaling Association, 22, 23, 96–99, 105, 107
pipelines: Dakota Access, 42; Enbridge, 42, 192–93; Keystone XL, 42

Prince of Whales, 5, 16, 22–23, 93, 94, 96–107, 207n20
Progress (ship), 23, 93, 107–14, *110*, 208n59, 209n73, 209n78, 211n59. *See also Charles Phelps* (ship); ship media
progress-based myth, 30, 208n50, 208n56. *See also* economic growth and capitalism

quaint, as term, 74, 205n32
quaintness: aesthetic category of, 70–78, 174, 205n32; exhaustion and, 80–86; in Nantucket tourism, 61–64, 205n44; people and, 78–80; postindustrial tourism and, 64–66; Sekula and, 87–92
Quakers, 54, 137, 181
Queequeg (character), 41, 43, 52, 54, 126. *See also Moby-Dick* (Melville)

racism, 123, 135–36, 186. *See also* whiteness; white supremacy and whaling heritage
railroad travel, 66–69, 94, 104–7
Random House, 157–58, *162*
rendering, 9–10, 22–23, 99. *See also* slaughtering
residual culture, 8–9. *See also* media studies; obsolescence
Riddell, John, 170
Rifkin, Mark, 17
Robinson, Ione, 180, 216n76
Rockefeller, John D., 2
rock oil. *See* petroleum
Rockwell Kent (Covarrubias), *172*
Rogers, H. H., 2
Rönnskog, Ann-Sofi, 192
Roosevelt, Theodore, 129
Rosaldo, Renato, 119, 149
Royal Navy, 17
R. R. Donnelley company, 157, 163
ruin pornography, 64–65, 89. *See also* infrastructure: ruins of
Ruskin, John, 163
Rydell, Robert, 112–13

San Clemente Dam removal, 193
Santa Barbara oil spill (1969), 18
Schultz, Elizabeth, 52, 158
Schumpeter, Joseph, 9
Sekula, Allan, 18, 20, 87–92; *Fish Story*, 18, 88–91
settler colonialism: economic power and, 87; fossil modernity and, 7, 16–17; glorification of, and nostalgia for, 23, 153; hunting and, 129; Indigenous nations and, 41–42, 201n45, 201n48; violence of, 43, 105, 120, 183–84; whale infrastructures and, 40–44, 111–13; whaling cultural narrative and, 210n8. *See also* white supremacy and whaling heritage
settler time, 17
Shakespeare, William, 102
Sharpe, Christina, 19, 91
shipbreaking, 184
ship media, 111–14. *See also Progress* (ship)
"ship of Theseus" paradox, 95
ships. *See* whaling ships; *and names of individual ships*
shipwrecks, 184, 194
Shoemaker, Nancy, 43
Shukin, Nicole, 9–10, 99, 101; *Animal Capital*, 99
Shulman, Peter, 148
Simpson, Mark, 13, 147
Sinclair Oil, 111
Sioux, 42
skeuomorphism, 165–66, *166*, 167–68, 215n43, 215n49. *See also* Kent, Rockwell
slaughtering, 33, 213n7. *See also* rendering
slavery, 19, 43
Sloan, John, 169
Smithson, Robert, 58
sport hunting. *See* hunting
S. S. *Constitution*, 185
Standard Oil company, 2, 134
Starbuck, Alexander, 75–76

Starosielski, Nicole, 14, 95
steamship travel, 66
Stieglitz, Alfred, 174
Stone Fleet, 107–8, 195n18
Stubb (character), 51. *See also Moby-Dick* (Melville)
Szeman, Imre, 13, 147

TallBear, Kim, 42
Tashtego (character), 41, 43, 126. *See also Moby-Dick* (Melville)
Technics and Civilization (Mumford), 154, 155
Territorial Agency, 192
textile industry, 189
38th Voyage, *Charles W. Morgan* (ship), 185–88, 194
Thunberg, Greta, 30
Tocqueville, Alexis de, 56
Todd, Zoe, 16, 42, 52
Tompkins, Kyla Wazana, 12
tooth, whale, 139, *141*. *See also* whale, as animal
tourism: Detroit ruin porn, 64–65; fossil fuels and, 66–70; quaintness in Nantucket, 61–64, 65–66
transatlantic slavery, 19, 43
Trump, Donald, 120, 147–48, 193–94
Turner, Frederick Jackson, 87

Understanding Media (McLuhan), 15, 99, 114

Viking Press, colophon for, *162*
Vineyard Wind, 191
violence: of extractive industries, 13, 18, 37, 200n17; of settler colonialism, 41–43, 105, 120, 183–84. *See also* rendering; slaughtering

"wake, the," 19
Wampanoag, 188
Wanderer (ship), 142–45, 194
Warren, Elizabeth, 26, 189–90
Weaver, Henry E., 108, 109, 111

Weaver, Raymond, 141; *Herman Melville: Mariner and Mystic*, 141, 150, 153-54
Wenzel, Jennifer, 4, 8, 33, 34
whale, as animal: baleen, 4, 5, 22, 44, 101; corpse, as entertainment, 5, 16, 22-23, 93, 94, 96-107, *107*, 207n20; jawbone, 40-41; tooth, 139, *141*; whalebone, 4, 122. *See also* rendering
whaleman statue (New Bedford, MA), 124-28, *125*
whale media, 95, 99-101. *See also* media, as concept
whale oil, 48-53; Black on, 195n16; extraction and, 32; magic, 34-35, 36, 44
whale tooth, 139, *141*
Whaling Enshrined, 183, 212n74
whaling entertainment: Prince of Whales corpse as, 5, 16, 22-23, 93, 94, 96-107, *107*; *Progress* (ship), 23, 93, 107-14, *110*, 211n59. *See also* whaling tourism
Whaling Film Corporation, 135, 138
whaling films, 135-42, 208n46
whaling industry, 1; as energy culture, 8-14; global rise of, 6, 213n7; Indigenous peoples and, 6, 42-44, 188, 201n46; Jones on, 94; labor and production in, 2, 34-35; in *Moby-Dick*, 29-36; settler colonialism and, 40-44; transition to petroleum and, 2-5; US Civil War and, 4-5
whaling infrastructures, 36-44, 111-13, 117-20, 203n94
Whaling National Historic Park, 189
whaling ships: *Charles Phelps*, 107; *Charles W. Morgan*, preservation of, 26, 142-47, 156, 183-91, 194, 212n71; deterioration of, 203n94, 203n96;

Pequod in Melville's *Moby-Dick*, 49, 54-55, 203n93; *Progress*, as entertainment, 93, 107-14, *110*, 208n59, 209n73, 209n78, 211n59; *Wanderer*, 143-44
whaling tourism, 5-6, 61-70. *See also* whaling entertainment
White, Stephen C., 184
whiteness, 84, 121, 125-27. *See also* racism; Yankee whaling heritage
white supremacy and whaling heritage, 23-24, 119-24, 130, 149. *See also* settler colonialism; Yankee whaling heritage
Whyte, Kyle Powys, 42
Wiindigo infrastructure, 40
Wilderness (Kent), 165, 166, *166*, 173
Williams, Raymond, 8
wind energy, 189-90, 191
Winfall, Silas, 68
Witch of Wall Street, 118
woodblock illustration style, 151; form, 157-63; method, 163-66, 179-80; politics of, 168-70
Workers of the World, Unite! (Kent), 163
World Made by Hand, The (Kunstler), 49
World's Columbian Exposition (1893), 23, 93, 97, 108-14, 211n59
Wyer, Henry Sherman, 78-80; *Nantucket Characters*, 78-80, *80*

Yaeger, Patricia, 161
Yankee Whaler, The (Ashley), 131, 146
Yankee whaling heritage, 24, 119-24, 149, 210n8. *See also* whiteness; white supremacy and whaling heritage
Yupik, 6

Zigrosser, Carl, 161

www.ingramcontent.com/pod-product-compliance
Lightning Source LLC
Chambersburg PA
CBHW032021230426
43671CB00005B/160